More praise for *Privacy Is Power*

'A forceful call for us to tame the data economy by reclaiming our privacy . . . and our power.'
Jonathan Zittrain, author of *The Future of the Internet*

'A bracing call to arms to fight back against digital surveillance before it is too late. If you're one of those readers who gave up before getting to the end of *The Age of Surveillance Capitalism*, Shoshana Zuboff's academic doorstopper, this is a good place to start.'
Richard Waters, *Financial Times*

'We didn't see digital surveillance coming, but today it's threatening democracy and basic freedoms. If you want to understand why privacy matters more than ever before, and how we can preserve it in an age of data grabbing, read this book.'
Nigel Warburton, author of *A Little History of Philosophy*

'In this bold, original, and engaging book, Carissa Véliz makes a compelling case for the revolutionary goal of reclaiming privacy from the grip of a destructive data economy. While many have lamented the ills of surveillance capitalism, Véliz's courageous manifesto paves a clear path for regaining power – taking back our ill-gotten information from tech companies and data brokers and reinvigorating democracy in the process.'
Evan Selinger, Professor of Philosophy, Rochester Institute of Technology and co-author of *The Cambridge Handbook of Consumer Privacy*

www.penguin.co.uk

'In this smart, stylishly written, and alarming volume
Carissa Véliz argues that it matters a great deal and that
we don't have to put up with it. Essential reading.'
Jonathan Wolff, author of *An Introduction to Moral Philosophy*

'A fantastic little book.'
Annabelle Lever, author of *On Privacy*

Privacy Is Power

Why and How You Should
Take Back Control of Your Data

CARISSA VÉLIZ

CORGI BOOKS

TRANSWORLD PUBLISHERS
Penguin Random House, One Embassy Gardens,
8 Viaduct Gardens, London SW11 7BW
www.penguin.co.uk

Transworld is part of the Penguin Random House group of companies
whose addresses can be found at global.penguinrandomhouse.com

First published in Great Britain in 2020 by Bantam Press
an imprint of Transworld Publishers
Corgi edition published 2021

A CIP catalogue record for this book
is available from the British Library.

ISBN 9780552177719

Typeset in 11/16pt Minion Pro by Jouve (UK), Milton Keynes
Printed and bound in Great Britain by Clays Ltd, Elcograf S.p.A.

The authorized representative in the EEA is Penguin Random House Ireland,
Morrison Chambers, 32 Nassau Street, Dublin D02 YH68.

Penguin Random House is committed to a sustainable future
for our business, our readers and our planet. This book is made
from Forest Stewardship Council® certified paper.

A mi madre, tierra firme que me dio alas

CONTENTS

INTRODUCTION

They are watching us. They know I'm writing these words. They know you are reading them. Governments and hundreds of corporations are spying on you and me, and everyone we know. Every minute of every day. They track and record all they can: our location, our communications, our internet searches, our biometric information, our social relations, our purchases, and much more. They want to know who we are, what we think, where we hurt. They want to predict and influence our behaviour. They have too much power. Their power stems from us, from you, from your data. It's time to take back control. Reclaiming privacy is the only way we can regain control of our lives and our societies.

The internet is primarily funded by the collection, analysis, and trade of data – the data economy. Much of that data is personal data – data about you. The trading of personal data as a business model is increasingly being exported to all institutions in society – the surveillance society, or surveillance capitalism.[1]

To reach you, I had to go through surveillance capitalism – I'm sorry.[2] How did you become aware of this book? Can you remember how you first heard about it, or where you saw an ad for it? You might've been tagged by some platform or another as a 'pioneer', someone who is on the lookout for knowledge and new experiences. You like books that make you think. Or you might be an 'advocate', someone worried about social issues and politically engaged. Fit the bill? The main objective of this book is to empower you, but most similar uses of your data will disempower you.

If surveillance didn't catch you before buying this book, it probably did afterwards. If you are reading these words on a Kindle, or Google Books, or a Nook, they are measuring how long it takes you to read each word, where you stop to take a break, and what you highlight. If you bought this book in a bookshop, the smartphone in your pocket was recording your journey there, and how long you stayed.[3] The music in the bookshop might have been sending ultrasound beacons to your phone to identify it as *your* phone and track your interests and purchases. If you used a debit or credit card to buy the book, they probably sold that information to data brokers who then sold it to insurance companies, possible employers, governments, businesses, and whoever else might have been interested in it. Or you may even have linked your payment card to a loyalty system, which tracks your purchasing history and uses that information to try to sell you more things the algorithm reckons you might buy.

The data economy, and the ubiquitous surveillance on which it feeds, took us by surprise. Tech companies did not inform users of how our data was being used, much less ask for our permission. They didn't ask our governments either. There were no laws to regulate the data trail left behind by unsuspecting citizens as we went about our business in an increasingly digital world. By the time we realized it was happening, the surveillance architecture was already in place. Much of our privacy was gone. In the wake of the coronavirus pandemic, privacy is facing new threats, as previously offline activities have moved online, and we have been asked to give up our personal data in the name of the public good. It is time to think very carefully about what sort of world we want to inhabit when the pandemic becomes a distant memory. A world without privacy is a dangerous one.

Privacy is about being able to keep certain intimate things to yourself – your thoughts, your experiences, your conversations, your plans. Human beings need privacy to be able to unwind from the burden of being with other people. We need privacy to explore new ideas freely, to make up our own minds. Privacy protects us from unwanted pressures and abuses of power. We need it to be autonomous individuals, and for democracies to function well we need citizens to be autonomous.

Our lives, translated into data, are the raw material of the surveillance economy. Our hopes, our fears, what we read, what we write, our relationships, our diseases, our mistakes, our purchases, our weaknesses, our faces, our voices – everything is used as fodder for data vultures who collect it all, analyse it all, and sell it to the highest bidder. Too many of those acquiring our data want it for nefarious purposes: to betray our secrets to insurance companies, employers, and governments; to sell us things it's not in our interest to buy; to pit us against each other in an effort to destroy our society from the inside; to disinform us and hijack our democracies. The surveillance society has transformed *citizens* into *users* and data *subjects*. Enough is enough. Those who have violated our right to privacy have abused our trust, and it's time to pull the plug on their source of power – our data.

It's too late to prevent the data economy from developing in the first place – but it's not too late to reclaim our privacy. Our civil liberties are at stake. The decisions we make about privacy today and in the coming years will shape the future of humanity for decades to come. Societal choices about privacy will influence how political campaigns are run, how corporations earn their keep, the power that governments and private businesses may wield, the advancement of medicine, the pursuit of public health goals, the risks we are exposed to, how we interact with each other, and, not least, whether our rights are respected as we go about our daily lives.

This book is about the state of privacy today, how the surveillance economy came about, why we should end the trade in personal data, and how to do it. Chapter One accompanies a person throughout a day in the surveillance society to illustrate how much privacy is being taken away from us. Chapter Two explains how the data economy developed, in the hope that understanding how we got into this mess will be helpful in getting us out of it. In Chapter Three I argue that privacy is a form of power, and that whoever has the most personal data will dominate society. If we give our data to companies, the wealthy will rule. If we give our data to governments, we will end up with some form of authoritarianism. Only if the people keep their data will society be free. Privacy matters because it gives power to the people.

The surveillance economy is not only bad because it creates and enhances undesirable power asymmetries. It is also dangerous because it trades in a toxic substance. Chapter Four examines why personal data is toxic and how it is poisoning our lives, our institutions, and our societies. We need to put a stop to the data economy like we put a stop to other kinds of economic exploitation in the past. Economic systems that depend on the violation of rights are unacceptable. Chapter Five is about how societies can pull the plug on the surveillance economy. Chapter Six is about what you can do as an individual to take back control of your personal data and our democracies.

We are not witnessing the death of privacy. Even though privacy is in distress, we are in a better place now to defend it than we have been for the past decade. This is only the beginning of the fight to safeguard personal data in the digital age. Too much is at stake to let privacy wither – our very way of life is at risk. Surveillance threatens freedom, equality, democracy, autonomy, creativity, and intimacy. We have been lied to time and again, and our data is being stolen to be used against us. No more. Having too little privacy is at odds with having well-functioning societies. Surveillance capitalism needs to go. It will take some time and effort, but we can and will reclaim privacy. Here's how.

ONE

DATA VULTURES

If you are reading this book, you probably already know your personal data is being collected, stored, and analysed. But are you aware of the full extent of privacy invasions into your life? Let's start at dawn.

What is the first thing you do when you wake up in the morning? You probably check your phone. *Voilà*: that is the first data point you lose in the day. By picking up your phone first thing in the morning you are informing a whole host of busybodies – your smartphone manufacturer, all those apps you have installed on your phone, and your mobile company, as well as intelligence agencies if you happen to be an 'interesting' person – what time you wake up, where you've been sleeping and with whom (assuming the person you share your bed with keeps his phone near him too).

If you happen to wear a smart watch on your wrist, then you will have lost some privacy even before you wake up, as it records your every movement in bed – including, of course, any sexual activity.[1] Suppose that your employer gave you that watch as part of a wellness programme to encourage healthy habits that can lead to cheaper insurance premiums. Can you be sure that your data will not be used against you? Are you confident your employer will not see it?[2] When your employer gives you a device, they remain its legal owner – be it a fitness tracker, a laptop, or a phone – and they can access the data from that device at any time without your permission.[3]

After checking your heart rate during the night (way too fast – you need to exercise more) and sending that data to your smartphone, you get out of bed and brush your teeth with an electric toothbrush. An app lets you know that you have not been brushing as often as you should.

You have overslept this morning, and your spouse has already left for work. You go to the kitchen and look for sugar for your coffee, only to realize you've run out. You decide to ask your neighbour if she has some to spare. Standing outside her door, you notice something unusual about it – there's a camera. When she opens the door, she explains: it's a new smart door-bell. If it's a doorbell from Ring, a company owned by Amazon, Ring employees will probably review that video footage of you in order to tag objects manually in an effort to train soft-ware to perform recognition tasks. Those videos are stored unencrypted, making them extremely vulnerable to hacking.[4] Amazon has filed a patent to use its facial recognition software on doorbells. Nest, owned by Google, already uses facial recognition in its cameras. In some cities like Washington DC, the police want to register, and even subsidize, private security cameras.[5] It's anybody's guess where footage from smart doorbells is going to end up and what it will be used for.

Your neighbour doesn't have sugar – or maybe she doesn't want to give you any after you sneered at her new doorbell. You are forced to settle for unsweetened coffee. You turn on the TV (a smart TV, of course) to distract yourself from its bitter taste. Your favourite TV show is on – that guilty pleasure you would never admit to watching.

You get a call. It's your spouse. You mute the TV.

'Why are you still at home?'

'How did you know?'

'My phone is connected to our smart meter. I could see you were using electricity.'

'I overslept,' you say.

He doesn't sound very convinced by your explanation, but he has an appointment and needs to end the call.

You wonder whether this is the first time you have been spied on through your smart meter. Smart meters are not only a privacy risk with respect to the people you share your home with. They are notoriously insecure devices.[6] A criminal can hack yours, and see when you're away from home, with a view to robbing your property.[7] Furthermore, data from smart meters is held and analysed by energy service providers. Some of this data can be quite sensitive. For example, your energy footprint is so precise that it can reveal which television channel you are watching.[8] The data can then be sold or shared with interested third parties.

Your teenage son suddenly walks in and interrupts your thoughts. He wants to talk to you about something. Something sensitive. Perhaps it's about a problem related to drugs, or sex, or bullying at school. You don't turn off the TV. It remains muted, playing images in the background. Your smart TV is probably collecting information through a technology called 'automatic content recognition' (ACR). It tries to identify everything you watch on TV, and sends the data to the TV maker, third parties, or both. Researchers found that one Samsung smart TV had connected to more than 700 distinct internet addresses after being used for fifteen minutes.[9]

That's the least of it. If you had time to read the privacy policies of the objects you buy, you would have noticed that your Samsung TV included the following warning: 'Please be aware that if your spoken words include personal or other sensitive information, that information will be among the data captured and transmitted to a third party'.[10] Even when you think you've turned your TV off, it might still be on. Intelligence agencies like the CIA and MI5 can make your TV look as though it is off while they record you.[11]

After your son has shared his most intimate thoughts with you, your TV manufacturer, and hundreds of unknown third parties, he leaves for school, where he will be forced to lose yet more of his privacy through the school's surveillance of his internet use.[12] You unmute the TV. Commercials are on. You think that you're finally going to get a moment of privacy.

You're wrong. Unbeknown to you, inaudible sound beacons are being broadcast through those TV (and radio) commercials (as well as through music in shops) and are being picked up by your phone. These audio beacons function like sound cookies that allow businesses to triangulate your devices and purchasing habits through location. That is, they help businesses track you across different devices. Thanks to this ultrasonic cross-device tracking, a company can find out whether the person who sees a particular ad for a product in the morning on TV, and looks it up on her laptop an hour later, then goes on to buy it at the shop in her neighbourhood, or orders it online.[13]

You receive another call. This time it's a colleague from work.

'Hey, I'm not sure how this happened, but I just received a recording of a very private conversation you were having with your son. It seems that your digital assistant Alexa sent it.'

You thank him for letting you know and hang up, wondering whether Alexa might have sent that conversation to other people in your contact list. Furious, you get in touch with Amazon. They explain: 'Echo probably woke up to a word in your conversation that sounded like "Alexa". Then, it thought you were saying "send message". It must have asked "To whom?" and whatever you were saying then was interpreted as a name.'[14] Sometimes smart speakers are activated by hearing a television show in which a word is spoken that is similar to their wake word. If you were to have your television on all day, that would happen between one and a half and nineteen times per day (not counting the times when the actual wake word is said on television).[15] When Alexa sent the private conversation of a user in Portland, Oregon, to a random contact, the user vowed never to plug in the device again.[16] You go a step further and smash the Echo against the wall. Your spouse will not be happy about that.

You are very late for work now. You get in your car and drive to your office. You bought your car second hand from an acquaintance. It probably never crossed your mind, but it turns out that person has access to your data, because she never disconnected her phone from the car app.[17] Furthermore, your carmaker is gathering all kinds of data about you – the locations you visit, the speed you drive at, your musical taste, your eye movements, whether your hands are on the wheel, and even your weight as measured by your seat – which can end up in the hands of your insurance company, among other third parties.[18]

You get to work. You live in London, and your office is in Westminster. As you pass the Houses of Parliament, data from your phone may be vacuumed up by IMSI-catchers, also called stingrays – fake cell phone towers that trick mobile phones into connecting to them. Once connected, IMSI-catchers collect identification and location data. They also allow eavesdropping on phone conversations, text messages, and web browsing.[19] According to the American Civil Liberties Union, at least seventy-five agencies in twenty-seven states have the technology in the United States (though there might be many more that we don't know about).[20] An article in the *Intercept* stated that law enforcement agents have sometimes 'deceived judges' and 'misled defense attorneys' about their use of stingrays, saying that they obtained information about the defendant from a 'confidential source', for example, when in fact they had used a stingray.[21] Activists believe that stingrays might have been used against Black Lives Matter protesters this year.[22] There is evidence that this equipment is being used by the police in London to spy on people, for example, at peaceful protests and near the UK Parliament.[23] Popular advice given online to protect your privacy includes leaving your phone at home when going to a protest, although having a phone is important during a protest, to keep up to date with relevant news and in touch with the people you know. Though stingrays are mostly used by governments, they can be used by anyone, since they are sold by private companies and can also be home-built.

While the data on your phone is being hoovered, you step into your office. A colleague greets you and looks at his watch, making it clear that your tardiness is being noted. You sit in front of your computer and try to inhale deeply, but you lose your breath at the sight of hundreds of unread emails.[24] You open the first one. It's from your boss. 'Hey, I noticed you weren't in the office this morning. Will you have that report I asked for ready on time?' Yes, you will, but you wish your boss weren't breathing down your neck.

The next email asks you to fill in anonymous evaluations of your co-workers. Your boss is a firm believer in work surveillance. You know that he tracks your every movement, keeping tabs on whether you attend meetings, seminars, and even informal dinners and drinks after work. You know he monitors your social media because he has warned you in the past about posting political content. You feel queasy at the thought of evaluating your co-workers and being evaluated by them.

Then there's an email from your favourite shoe brand. You might think it harmless to your privacy to receive emails, but about 70 per cent of commercial emails and 40 per cent of all emails contain trackers.[25] Opening an email allows third parties to track you across the web and identify you as one and the same user across devices. Trackers can be embedded in a colour, a font, a pixel, or a link. Even ordinary people are using trackers to know whether their emails are being read, when, and where. Given that trackers can reveal a person's location, stalkers can use them to find you.

The next email is from your brother. He has used your work address even though you have asked him not to. Employers, including universities, have access to your emails[26] – one reason among others never to use your work email for personal purposes. In his email, your brother informs you that he received a direct-to-consumer genetic testing kit as a birthday present and went ahead and tested himself. You might be happy to learn, he writes, that our family is 25 per cent Italian. The bad news is that he has a 30 per cent chance of getting heart disease; given that he is your brother, that probably applies to you too. You reply: 'I wish you had asked for my consent. Those are my genes too, and those of my child. Didn't you know our grandmother was Italian? If you want to know more about our family, ask me.'

Worried about your genetic data, you look at the privacy

policy of the company your brother used. It doesn't look good. Testing companies can claim ownership of the DNA sample you send them, and use it in whichever way they want.[27] Privacy policies of DNA testing companies usually refer to the 'de-identification' or 'pseudonymization' of information to reassure people. Genetic data is not easy to de-identify effectively, however. It is in the nature of genetic data that it can uniquely identify individuals and their family connections. Replacing names with randomly generated ID numbers does not provide much security against re-identification. In 2000, computer scientists Bradley Malin and Latanya Sweeney used publicly available healthcare data and knowledge about particular diseases to re-identify 98 to 100 per cent of individuals on an 'anonymized' DNA database.[28]

You wonder what will become of your brother's genetic data, and whether it will ever count against you or your son if you apply for insurance or a job, for example. The worst part of it is that home genetic tests are incredibly imprecise. About 40 per cent of results are false positive.[29] Your brother may have given away all of the family's genetic privacy in exchange for mumbo jumbo that will nonetheless be treated as facts by insurance companies and others.

It's time for a work videoconference with a client, who has requested you connect through Zoom. Many people hadn't heard about Zoom before the coronavirus pandemic, when it became the most popular videoconferencing app. During lockdown, you were horrified to learn of the reams of data Zoom collected on you, including your name, physical location, email address, job title, employer, IP address, and much more.[30] You have a vague idea that Zoom has improved its privacy and security policies now, but can you trust a company that claimed to implement end-to-end encryption when it didn't?[31]

When the call is over, in an effort to relax you log on to Facebook. Just for a little while, you tell yourself. Perhaps the photos of your friends' good times will cheer you up (they won't). Given that you suspect your boss monitors what you do on your computer, you use your personal phone.

Facebook has violated our right to privacy so many times that a comprehensive account would merit a book in itself. Here I mention only a few ways in which it invades our privacy.

Everything you do while on Facebook gets tracked, from your mouse movements[32] to the things you write and decide to delete before posting (your self-censorship).[33] You start browsing the section entitled 'People You May Know'. This feature has been crucial in expanding Facebook's social network, which went from 100 million members when the tool came out in 2008 to over 2 billion in 2018. Among the people you can see there you may recognize distant relatives, or people you went to school with. That doesn't sound too bad. I suggest you don't go much deeper into that rabbit hole though. If you do, you may likely find that Facebook is trying to connect you with people you do not want to connect with.

Some connections between people are problematic, like when the real identities of sex workers are outed to their clients.[34] Or when a psychiatrist's patients get linked together, throwing medical confidentiality out of the window. The psychiatrist in question was not friends with her patients on Facebook, but her patients probably had her in their contact books.[35] Among many other ill-fated connections, Facebook has also suggested as friends a harasser to his (previously anonymous) victim, a husband to his wife's lover, and a victim to the man who robbed her car.[36]

Facebook's current mission statement is to 'give people the power to build community and bring the world closer together'. What about giving people the power to *disconnect* from toxic or undesirable relationships? 'Bringing the world closer together' sounds cosy until you ask yourself whether you want closeness forced upon you with people you fear, dislike, or want to keep at a distance for professional or personal reasons.

Facebook has proved its lack of respect for privacy in many other ways. About 87 million Facebook users had their data analysed for political purposes by the data firm Cambridge Analytica.[37] In 2018, 14 million accounts had personal data stolen in a hack.[38] For years, Facebook allowed Microsoft's Bing search engine to see Facebook users' friends without their consent, and it gave Netflix and Spotify the ability to read and even delete Facebook users' 'private' messages.[39] In 2015, it started logging all text messages and calls from Android users without asking for their permission.[40]

Facebook has probably used facial recognition on your photos without securing proper consent from you. When Tag Suggestions asked you 'Is this Jack?' and you responded 'yes', what you did was give away your friend's privacy and your labour for free to train Facebook's facial recognition algorithm. Facebook has filed patents that describe systems to recognize shoppers' faces in stores and match them to their social networking profiles.[41] To top it off, Facebook asked users for their telephone numbers as a security measure, and then used that information for its own purposes – to target ads and unify its data sets with WhatsApp, its messaging app.[42] In 2019, hundreds of millions of Facebook users' phone numbers were exposed in an open online database, because the server they were on was not password-protected.[43] These are just some of the latest disasters, but the full list is a long one, and everything seems to indicate that Facebook's violations of our right to privacy are not about to stop.[44]

Facebook may seem like a social network on its surface, but its real business is trading in influence through personal data. It is more of a personalized advertisement platform than a social medium. It is willing to go to great lengths to scrape up as much personal data as possible with as little friction as possible so that it can sell advertisers access to your attention. Judging by its history, if it can get away with it – and so far it has – Facebook will not ask for consent, it will not make an effort to investigate who is getting your data and how it is being used, and it is willing to break its promises.[45] Protecting your privacy seems to be the lowest priority on its list. And you can't even stay away from this data-hungry monster, because Facebook has a shadow profile on you even if you are not a user. It follows you around the web through its pervasive Facebook 'like' buttons, even if you don't click on them.[46] It's no wonder a British parliamentary report has suggested that Facebook has behaved like a 'digital gangster' in the past few years.[47]

After browsing Facebook for a while and feeling 'creeped out' about the friends it suggests and the ads it shows you, you take a break from it. You try to get down to some work, but you cannot concentrate, thinking about how your boss is likely monitoring every move you make on your computer. Luckily for you, it's time for lunch. Except you're not hungry, so you decide to go to a nearby shop to buy something for your son that may help him feel better.

You go into a clothing shop to find a shirt. Bricks-and-mortar businesses have felt disadvantaged in comparison to online shops because the latter were the first to collect oceans of data from customers. So physical shops are trying to catch up. The shop you enter uses technology that identifies you as a returning shopper through your Wi-Fi mobile signal. Mobile devices send unique identification codes (called MAC addresses) when they search for networks to go online. Shops use that information to study your behaviour.[48]

Not content with that, shops may also use cameras to collect data on you. Cameras may help to map customers' paths and study what people are attracted by, how they navigate the shop. Cameras have become so sophisticated that they can analyse what you are looking at and even what mood you're in based on your body language and facial expression.[49] The shop may also be using facial recognition. Among other uses, facial recognition allows businesses to cross-reference your face with a database that looks for a match with past shoplifters or known criminals.[50]

You step out of the shop and check your phone. An alert

reminds you that you have a doctor's appointment. There's a health issue that has been bothering you for some weeks. You've searched online, trying to find a solution, and hoped it might go away on its own, but it hasn't. You haven't told anyone in your family so as not to cause unnecessary worry. Our search engines know more about us than our spouses: we never lie to them or conceal our worries from them.

You go to the doctor. You receive a notification while you are in the waiting room. Your sister has posted the latest photo of your baby niece. Her chubby hands make you smile. You make a mental note to warn your sister about the dangers of exposing her kids online. You ought to tell her that our online photographs are being used to train facial recognition algorithms that are then used for all sorts of nefarious purposes, from the surveillance of vulnerable populations by authoritarian regimes to outing pornography actors and identifying strangers on the subway in Russia.[51] But your niece's irresistible smile distracts you. Pictures of her are sometimes the highlight of your day, and the kind of thing that makes the data economy bearable, even when you know that engaging content such as endearing babies is precisely what the data vultures feed on.

A nurse announces that the doctor is ready to see you. As your doctor asks sensitive questions, records your answers on her computer, and schedules some tests for you, you wonder where that data might end up. Your medical data is often for sale. Data brokers[52] – traders in personal data – can acquire medical data from pharmacies, hospitals, doctors' offices, health apps, and internet searches, among other sources. Your medical details

could also end up in the hands of researchers, insurance companies, or prospective employers.[53] A health system like the NHS
(the UK's National Health Service) may decide to donate your
records to a company such as DeepMind, owned by Alphabet
(Google's parent company). The transfer of data may be done
without your consent, without you benefiting from such an
invasion of privacy, and without any legal guarantee that Deep-
Mind will not link your data to your Google account, thereby
further eroding your privacy.[54] In 2019, the University of Chicago
and Google were sued. The potential class-action lawsuit accused
the hospital of sharing hundreds of thousands of patients'
records with Google without deleting identifiable date stamps or
doctors' notes. It also accused Google of 'unjust enrichment'.[55]

You could also experience a data breach. In 2015, over
112 million health records were breached in the United States
alone.[56] You could even become a victim of extortion. In 2017,
criminals got access to medical records from a clinic and blackmailed patients; they ended up publishing thousands of private
photos, including nude ones, and personal data including passport scans and national insurance numbers.[57]

As these thoughts race through your mind, you feel tempted
to lie to your doctor about sensitive information that (you hope)
is not necessary to get an accurate diagnosis. You may even feel
encouraged not to get the prescribed tests done at all, even though
you need them.

After the doctor's, you go back home to pack for your work
trip to the United States. All day you have been tracked by the
apps on your phone. If you allow location services to be on so

you can receive local news, the local weather, or other similar information, dozens of companies receive location data from you. In some cases, these apps update and receive your location data more than 14,000 times a day. Location-targeted advertising is a business worth an estimated $21 billion.[58]

Among the many hands that are trading your location data around are telecoms. Jealous of Silicon Valley's business success, telecoms are eager to compete in the data trading market.[59] Your mobile phone is constantly connecting to the nearest cell phone tower. As a result, your mobile service provider always knows where you are. Mobile networks not only sell location data to other companies; journalists have exposed that at least some mobile service providers are also selling your data on the black market. The upshot is that anyone with a mobile phone is vulnerable to being watched by stalkers, criminals, low-level enforcement officers who do not have a warrant, and other curious third parties who may have very questionable intentions and no right to access our sensitive data. In the United States, obtaining real-time updates on any mobile phone's location costs around $12.95.[60] While this underground market for location data has only been confirmed in the United States with respect to T-Mobile, Sprint, and AT&T, it may well be happening with other telecoms and in other parts of the world.

Car companies, data brokers, telecoms, shops, and tech giants all want to know where you are. You might find it reassuring to think that, even if it's true that huge amounts of data about you are being collected, much of it will be anonymized. Unfortunately, all too often it is easy to re-identify anonymized data. One of the first lessons in re-identification came from Latanya Sweeney in 1996, when the Massachusetts Group Insurance Commission published anonymized data showing the hospital visits of state employees. Then Governor William Weld assured the public that patients' privacy was safe. Sweeney proved him wrong by finding his medical records in the data set and mailing them to his office. She later showed that 87 per cent of Americans could be identified with three pieces of data: birth date, gender, and zip code.[61]

Another way in which you could be identified is through your location. Every person has a different location footprint, so even if your name is not on the database, it's easy to work out who you are. The specificity of location data is not surprising, given that there is usually only one person who lives and works where you do. Yves-Alexandre de Montjoye, head of the Computational Privacy Group at Imperial College London, studied fifteen months of location data for 1.5 million individuals. De Montjoye and his colleagues found that, in a data set in which people's locations are recorded on an hourly basis with a spatial resolution equal to that given by mobile phones as they connect to cell towers, it is enough to have four spatio-temporal data points to uniquely identify 95 per cent of individuals.[62] Similarly, when researchers looked at three months of credit card records for over a million people, they found that they needed only four spatiotemporal data points to uniquely re-identify 90 per cent of individuals.[63]

Databases can often be de-anonymized by matching them

with publicly available information. In 2006, Netflix published 10 million movie rankings by half a million customers as part of a challenge for people to design a better recommendation algorithm. The data was supposed to be anonymous, but researchers at the University of Texas at Austin proved they could re-identify people by comparing rankings and timestamps with public information in the Internet Movie Database (IMDb). In other words, if you saw a movie on a particular night, liked it on Netflix, and then rated it on IMDb as well, researchers could infer that was you. Movie preferences are sensitive; they can reveal political and sexual tendencies. A lesbian mother sued Netflix for putting her at risk of being outed.[64]

Data brokers are misleading the public when they claim they anonymize data.[65] They trade in personal data. They collect all kinds of extremely sensitive information, package it, and sell it to banks, insurers, retailers, telecoms, media companies, governments, and occasionally, criminals.[66] These companies sell information about how much money you make, whether you are pregnant, or divorced, or trying to lose weight. They have also been known to sell lists of rape victims, AIDS patients, and other problematic categories.[67]

Online ads also use questionable categories to target individuals. The Interactive Advertising Bureau, a trade group that establishes industry norms, uses categories for targeted ads that include incest or abuse support, substance abuse, and AIDS/HIV. Google's categories to target people for ads likewise include substance abuse, sexually transmitted diseases, male impotence, and political leanings.[68] These categories show what data vultures are interested in: they are eager to know where we hurt the most. Like predators, they can smell blood. They look for our vulnerabilities in order to exploit them.

Let's go back to your day. We left you packing for your work trip to the United States. When you get to Heathrow airport, you might not be asked for your boarding pass as you go through security and later board the plane. Facial recognition is now being used to verify your identity.[69] In the United States, airlines like JetBlue and Delta are already using this technology. And President Trump issued an executive order mandating the use of facial recognition identification for '100 percent of all international passengers', including American citizens, in the top 20 US airports by 2021 (even though there seem to be doubts over its legality).[70]

Back to your trip. When you arrive at your destination, a Transportation Security Administration (TSA) officer asks you to surrender your laptop and smartphone. You try to resist, but he informs you that if you deny his request, you will be refused entry. You have a work event to get to. If your boss learns that you did not attend that meeting because you were deported for disobeying an officer at border control, he will not be pleased, to put it mildly. You wonder whether he might even fire you. The thought of being unemployed motivates you to surrender your most private data. You try to think about the kind of data you have in there. You think about the nude photos with your spouse, photos of your children, all your financial information.

Then it occurs to you that you also have very private information regarding your employer. Perhaps you have business secrets that are worth millions. How can you be sure that data will not end up in the hands of an American competitor? Or perhaps you have confidential information about your government that you produced or acquired when you worked as a consultant. In 2017, a NASA engineer was forced to unlock his smartphone at the border, even though it had sensitive content on it.[71] Or maybe you are a doctor who has sensitive information about your patients on your laptop, or a lawyer protecting your clients, or a journalist protecting your sources.

You tell the TSA officer that you have to protect whatever confidential information you have – it is your professional duty, and you could face legal consequences if you do not. The TSA officer is unmoved. You remember reading something in the press about how, if you get deported from a country, you have to stay away for five or ten years. That would be fatal for your job. You're not sure whether 'refused entry' equals deportation. You ask for a lawyer. The TSA officer responds by saying that, if you want a lawyer, you must be a criminal. He asks whether you have something to hide. Weary and intimidated, you end up complying and handing over your laptop and phone. He takes your electronic devices away from your sight for a quarter of an hour or more. In that time, he downloads your data.[72]

Smart borders are becoming threats to civil liberties; they are being rolled out without seriously evaluating their benefits, risks, and legal and ethical implications.[73] Drones, sensors, and facial recognition, among other invasive technologies, promise cheaper and more effective border control at the cost of our privacy. Given its failure to fund a brick wall at the border with Mexico, the Trump administration built a virtual wall made of surveillance. Sensors are not only deployed at the actual border, but also in American communities close to the border.[74] Similar initiatives are being proposed and tested around the world. Hungary, Latvia, and Greece piloted an automated lie-detection test at four border points. The system, called iBorderCtrl, asked travellers questions such as 'What's in your suitcase?' and then tried to identify 'biomarkers of deceit'.[75]

You arrive at your hotel feeling exhausted, angry, and humiliated by the violation of your right to privacy. You resolve to do something to avoid or minimize future violations. You think about writing an email to an immigration lawyer, to be better informed about your rights. But you are afraid that the TSA, the NSA (National Security Agency), or some other agency might get access to that message, and that that might be enough to get flagged at airports. You don't want to become that person who always gets stopped at borders and interrogated for hours at a time. You're too afraid to ask for legal advice. Maybe it's enough if you lessen the data your phone and laptop collect on you. It's a start, in any case.

You might begin by trying to establish what data may have been downloaded from your phone and laptop. You download the data that Google and Facebook have on you.[76] Horrified at the level of intrusion you discover (Google has data on you that you thought you had deleted), you decide you should change all your privacy settings to minimize data collection. When you look at your settings, you notice that all the defaults are set to undermine your privacy.[77] And, while some of these settings can be changed, if you do not consent to some data collection, you cannot use the services provided by tech giants like Facebook and Google.[78] There is no room for negotiating the terms and conditions, and these can change at any time without you being notified.[79] You are being bullied.[80]

It dawns on you that in many ways you are being treated as a criminal suspect – the level of intrusion, the geotracking as if you had an electronic bracelet attached to your ankle, and the forcefulness of it all. In some ways it's worse than being a suspected criminal. At least when the police arrest you they allow you to remain silent, and warn you that anything you say may be used against you. As a subject of tech, you have no right to remain silent – trackers collect your data regardless of your not wanting them to – and you are not reminded that your data can and will be used against you. And at least during a trial you wouldn't be forced to self-incriminate. In the surveillance society, your data is used against you all the time.

Your spouse interrupts your thoughts with a call. He is upset about the smashed Echo. Things haven't been great between you for a while. You wish you had the serenity to tell him calmly what happened, but you feel defeated. Your silence causes an escalation in the distress your partner feels and expresses. 'I'm sorry,' he says, 'I wish we were face to face, but I cannot stand a day longer in this situation. I would like a divorce. We'll talk about the details when you come back.' He hangs up on you.

Stunned, you open Spotify on your laptop to calm yourself

down with music. The first ad that comes up is for a divorce lawyer. Is it a coincidence? Probably not. How did they know? And who are 'they'? Maybe it was your spouse's online searches for divorce. Or perhaps it was your marital fights that got recorded and analysed. Or maybe a predictive algorithm guessed your impending divorce based on how little time you have spent with your family lately. Maybe it was Spotify analysing your mood based on your music choice. Even bankers are gauging the public mood by looking at data from Spotify.[81] It bothers you that you'll probably never know who knows that you are getting a divorce, how they got that information, and whether they knew before you did. Whatever the case, it is not okay. You didn't tell them, and they do not have a right to eavesdrop on your most intimate relationships.

You wonder how far the privacy invasions can go before we decide to limit them. Technology has always pushed the boundaries of privacy. First photography, now the internet. You shudder at the memory of the news that Nike has started to sell their first smart shoes.[82] If researchers develop 'smart dust' – ubiquitous sensors that don't need batteries and are tiny enough to be near invisible[83] – privacy might become almost impossible to protect.

You are tempted to think that you'll be glad to leave this brave new world behind, one day. You're only sorry that your son has had to cope with privacy problems from such a young age, and that he will have to deal with them for much longer than you. Pondering your mortality, it dawns on you that violations of your right to privacy will not stop with your death. You will keep on living online. Scavengers will keep living off the trail of data you leave behind. And perhaps that data may still affect your son and his offspring. It may also affect the way your life is perceived by others – your post-mortem reputation.

You wonder if there is anything you can do to clean up your data footprint before it's too late. There is. Before yielding to despair over how we lose streams of privacy every second of every day, read on. The next three chapters do not paint a pretty picture, but taking a look at the grisly innards of the data economy is important to allow us to better understand how we got here, and how we can get out of this oppressive mess.

TWO

HOW DID WE GET HERE?

The contrast between today's privacy landscape and that of the 1990s is stark. At the end of the twentieth century, your car was a car – it wasn't interested in the music you like, it didn't listen to your conversations, it didn't track your weight, it didn't record your comings and goings. Your car took you where you wanted to go. It served you. You didn't serve it. For some of us, waking up to surveillance in the digital age felt as though we went to bed one night and found a completely different world the next morning – a bleaker one, at least with regard to our privacy and our autonomy over the objects surrounding us. How did we get here? Why did we allow the surveillance society to take root? At least three elements played a part in the erosion of our privacy: the discovery that personal data result-ing from our digital lives could be very profitable, the terrorist attacks of 11 September 2001, and the mistaken belief that privacy was an outdated value.

TURNING DATA EXHAUST INTO GOLD DUST

How do your everyday life experiences end up being data? Through your interaction with computers. Computing produces data as a by-product. By using digital technologies, or by digital technologies using you, a data trail is created of what you've done, when, and where. At the beginning of the digital age this data was not used commercially – either it was not used at all, or it was used as feedback to improve the system for users. The main protagonist in the story of the transformation of data exhaust into data gold dust is Google.[1]

Larry Page and Sergey Brin met at Stanford University in 1995 when they were both students. In 1996 they developed the core of Google – the PageRank algorithm.[2] PageRank counts the number and quality of links to a page to assess how authoritative that website is, and it ranks search results accordingly. The algorithm assumes that the more important websites are those that receive the most links from other authoritative websites. Other search engines returned irrelevant listings because they only focused on the text, without giving different weight to different kinds of sources. Page and Brin's algorithm, in contrast, could give more visibility to newspapers than to unknown blogs, for instance.

PageRank was inspired by academic citations. Academics

write papers building on other people's work, which they cite in their own work. The more citations a paper receives, the more important it's considered to be. By imitating academia, Page-Rank managed to establish order out of the meaningless noise of the internet, and thus make search queries much more informative and valuable. It was a brilliant idea. Not only that, the algorithm got better and better as the internet grew. It scaled spectacularly well.[3]

Unfortunately for us all, the trouble was that Page and Brin wanted to turn Google Search from an amazing tool into a money-making company. Early in 1999 they tried to sell Google to Excite, another search engine, but were unsuccessful. They also reportedly tried to sell it to AltaVista and Yahoo.[4] In 2000, two years after it was incorporated and despite its growing popularity, Google still hadn't developed a sustainable business model. In that sense, it was just another unprofitable internet start-up. Investors were growing impatient. One of them joked that the only thing he had received from his six-figure investment was 'the world's most expensive T-shirt'.[5] There was a risk of funders pulling out if the company didn't start earning money, fast. Google's financial situation was desperate.

The tide turned quickly. In 2001, Google's revenue increased to $86 million from $19 million in 2000. In 2002, that figure jumped to $440 million, then $1.5 billion in 2003, and $3.2 billion in 2004. That's a 3,590 per cent increase in revenue in only four years, from 2001 to the end of 2004.[6] How did they do it? No, they didn't rob a bank or find oil beneath their feet – not quite. They used the personal data of their users to sell ads, thus inaugurating the age of 'surveillance capitalism', as social psychologist Shoshana Zuboff has brilliantly dubbed it.

Before they became the world masters of ads, people at Google had not looked favourably on advertising. Or so they claimed. Brin and Page wrote a paper in 1998 in which they expressed their concerns about depending on ads. 'We expect that advertising funded search engines will be inherently biased towards the advertisers and away from the needs of the consumers,' they wrote. The paper seemed to suggest they wanted to keep Google as an academic tool: 'we believe the issue of advertising causes enough mixed incentives that it is crucial to have a competitive search engine that is transparent and in the academic realm'.[7] It is too bad things turned out so differently. PageRank was more trustworthy than previous search engines precisely because it didn't rely on advertising to make a profit, and therefore it didn't need to bias its results. Based on their paper, Brin and Page seemed unlikely candidates to turn the internet into an advertising marketplace.

Eric Veach was the person who engineered the ads system that made Google what it is today. 'I hate ads,' he said, echoing Brin and Page's stance.[8] To their credit, it seems that Googlers intended to make ads better than the average online ad in those days. The theory of AdWords, as the system was called, sounds reasonable enough: ads should make Google, advertisers, and users happy. Google gets money, advertisers get to publicize and sell their products, and users get to use a high-quality search engine while being exposed to ads that interest them. Doesn't sound like too bad a deal.

One special feature of AdWords was that it wasn't up to advertisers to buy the best positions. Rather, ads that managed to get more people to click on them were prioritized, which was meant to ensure that ads were useful to users. The system was easy to game, however: advertisers could click on their own ads to gain visibility, which is why Google decided to start auctioning ads instead. Advertisers would pay by the click. That is, advertisers would submit a bid for how much they were willing to pay for each time a user clicked on their ad. The advertiser who won the bid was to pay one penny more than the runner-up. It's a clever system, and it was a game changer. By charging by the click, Google allowed advertisers to pay for ads only when they worked. A further innovation was that Google lowered the price for more effective ads, thereby incentivizing better ads.

There was much to like about Google's ads, compared with others. They were relatively unobtrusive, they were clearly marked as 'sponsored' and not mixed in with a user's 'organic' searches, and there were incentives for quality. But there were less palatable aspects to the system too. One was that it created a 'black box' for advertisers, who had to trust Google's calculations and could never get the full story behind how Google placed their ads and why.[9] A greater downside of the Google ads system was that it turned the company's business model upside down. Google's users ceased to be its clients; its customers were now the advertisers. And we, the users, became the product. Google's incentives and loyalties had taken a drastic turn.

To this day, Google is primarily an advertising company. It earned almost $135 billion in 2019 just from ads. Alphabet,

Google's holding company, had total revenues of almost $162 billion that year. In other words, more than 80 per cent of Alphabet's returns are from Google ads.[10] And AdWords is still the most profitable of Google's advertisement initiatives.[11]

The most troubling victim of Google's advertising success, as you can imagine, was our privacy. Our data, which up until then had only been used to improve Google's search engine, began to be used to personalize ads. Through our searches, Google built a precise image of our minds, both collectively and individually.

We tend to search for what we're thinking about. On 28 February 2001 there was an earthquake near Seattle at 10.54 in the morning. Google had learned about it by 10.56 after detecting a surge in earthquake-related searches in that area.[12] It also knows what TV shows are most popular at any one time. And it has access to even more private information, such as whether you are thinking about doing drugs, or having an abortion, or whether you're concerned about your health, or about not being able to pay back a loan. A live sample of all these searches was shown at Google through a display called Live Query. A *New York Times* reporter wrote that looking at Live Query felt like 'watching the collective consciousness of the world stream by'.[13]

All that data can be used to sell ads. By 2003, the concept was well developed enough that Google's computer scientists filed a patent entitled 'Generating User Information for Use in Targeted Advertising'.[14] Patents are a good way to know what companies are up to. This patent not only described how to target ads with data that users left behind as they searched through Google, it also described how to infer data that users might not provide 'voluntarily'. What the patent illustrates, among other things, is that Google went from receiving data produced by users interacting with its website and using it to improve its service, to *creating* and *hunting* for user data with the express objective of using it to target advertisements.

As users searched for what they desired, feared, and were curious about, Google collected oceans of data on them. But once users clicked on a search result and left for another website, they were beyond Google's reach. That is, until the tech giant designed AdSense to complement its AdWords. AdSense uses the internet as if it were a blank canvas ready to be plastered with its ads – it's almost everywhere around the web. It runs ads in websites that are (or were) independent from Google, such as online shops and newspaper websites. With AdWords and AdSense, Google had kicked off the surveillance economy.

Before Google, some personal data was bought and sold here and there. And some of it was used for advertising, but not on a grand scale. Not with that level of specificity and analysis. Not for the purposes of personalization. Not as the main funding scheme for much of the internet. Google successfully turned data exhaust into gold dust and inaugurated the surveillance economy as one of the most lucrative business models of all time. It mixed all of the existing ingredients, and then some, and baked the cake. A race to the moral bottom developed as other companies tried to catch up, developing their own ways of mining our personal data. Google had turned its users into products, and others followed.

In an effort to continue its surveillance supremacy, in 2007

Google bought DoubleClick, an ad company that used a 'cookie' (a small piece of data that identifies visitors to websites) to access users' personal data, including browsing history, before the user even clicked on an ad. DoubleClick also used display ads (graphic banners), which further violated Google's original position of not creating distracting ads. Thanks to DoubleClick, Google could now follow users almost everywhere they went online, even if they didn't click on any ads.[15] Since then, Google has created product after product to help it gather even more data from even more sources. Chrome, Maps, Pixel, Nest, and many others were designed as ways to collect even more data from you. Why would a company offer an extra service like Maps, something that is so effortful to create and maintain, for nothing in return? It wouldn't. Google wanted to mine your location data.

Most of us learned about the dubious practices behind what we took to be our tech heroes very late in the game. Google and other businesses started profiting from the collection, analysis, and trade of our personal data without asking permission from governments or securing consent from users. They just went ahead with their plans with a 'let's see what happens' attitude. And nothing happened. Spellbound by the novel and 'free' services offered by tech companies, users accepted what we thought was a bargain without realizing what we were giving up.

The first time you opened an email account it probably never occurred to you that you were surrendering your personal data in return. It certainly didn't cross my mind. It wasn't a fair and open deal. The narrative about users consciously exchanging personal data for services was told to us years after the deal had been sealed, once tech companies had our data and getting out of the transaction seemed impossible.

Never was it more obvious that our interaction with digital technologies is not entirely voluntary than during the coronavirus lockdown. People were forced to use technologies that are extremely disrespectful of privacy, such as Zoom, for their work, to keep their children in school, and to keep in touch with their family. Once digital platforms became indispensable to us, a must-have in order to be a full participant in our society, there was no chance of opting out of data collection.

Make no mistake about it: it is no coincidence that you learned about surveillance capitalism long after it became mainstream. Google kept very quiet about its personal data collection endeavours and its business model.[16] The then CEO Eric Schmidt called it 'the hiding strategy'.[17] Secrecy was a way to protect their competitive advantage for as long as possible. It was also a way to keep users in the dark about what was being done with their data. As former Google executive Douglas Edwards wrote: 'Larry [Page] opposed any path that would reveal our technological secrets or stir the privacy pot and endanger our ability to gather data. People didn't know how much data we collected, but we were not doing anything evil with it, so why begin a conversation that would just confuse and concern everyone?'[18] As we will see, we should worry about our private data being collected even if no one is using it at the moment for evil purposes – data often gets misused eventually. Furthermore, what is evil is not always apparent, especially when it becomes part of a system. The data economy led, years later, to the erosion of equality, fairness, and democracy. That's why the right to privacy is indeed a right, and we should respect it even when the bad effects are not immediately obvious.

Google kept its mouth shut about its business model because it was taking something very private from us without asking for our permission and using it for the benefit of its advertisers and itself.[19] And no one stopped it because almost no one knew what was going on. But things might have been different. The surveillance economy was not inevitable. Brin and Page could have become academic professors, and kept Google Search as a non-commercial academic initiative – something like Wikipedia. Or they could have found an alternative business model. Or regulatory bodies could have limited what could be done with our private data. In fact, the data economy was almost regulated, but disaster got in the way.

DISASTER STRIKES

By the late 1990s, regulatory bodies were getting worried about cookies. In 1996 and 1997, Federal Trade Commission (FTC) workshops in the United States discussed giving people control over their personal information. The FTC's first approach was to encourage companies to self-regulate. But companies didn't listen. In 1999, for instance, DoubleClick had merged with a data broker, Abacus, presumably to try to identify its users. Privacy advocates petitioned the FTC to investigate, and DoubleClick was pressured to sell Abacus.[20]

When it became clear that self-regulation wasn't enough to

protect users' privacy, the FTC took a further step. In 2000, it wrote a report to Congress proposing legislation. It suggested that websites inform users of their information practices, that users be given choices as to how their personal data was used, that websites allow people to access the personal data held about them, and that websites be required to protect the security of the information they collect. 'The Commission believes that industry's limited success in implementing fair information practices online, as well as ongoing consumer concerns about Internet privacy, make this the appropriate time for legislative action,' reads the report.[21] If the United States had legislated to curb personal data collection online then, our world might look very different today. Google might never have become an advertising giant, and the surveillance practices that have become commonplace might never have developed.

Regrettably, history unfolded in a very different way. A little more than a year after the FTC report came out, in September 2001, four passenger airplanes in the United States were hijacked by terrorists. Two crashed into the Twin Towers in New York City, one crashed into the Pentagon, and the fourth one, apparently heading for the White House, crashed in Pennsylvania after its passengers fought the hijackers. The attack not only claimed the lives of almost 3,000 people, it unleashed war, was used to justify the passing of extraordinary laws, and inflicted a national and international trauma the effects of which linger to this day. Tragically, the terrorist attacks were extremely successful in wounding liberal democracy. And part of the damage was inflicted by elected representatives.

After 9/11, the mandate pronounced by President George W. Bush and echoed in American society for years was 'never again'. There was a feeling of shame for not having prevented the attacks, and a resolve to do whatever was necessary to make sure there could never be another 9/11. Overnight, the focus of government action in the United States became security. Privacy regulation got shelved.[22] It wasn't only a matter of the government being too busy with security to tackle privacy. Intelligence agencies saw an opportunity to expand their surveillance powers by obtaining a copy of all the personal data that corporations were collecting.[23] Once the government took an interest in our personal data, there was no incentive for them to regulate for privacy. Quite the contrary: the more data businesses collected, the more powerful government surveillance could be, and the more terrorist attacks could be prevented – in theory.

The United States Congress passed the Patriot Act, instituted a Terrorist Screening Program, and created a number of other measures that increased warrantless surveillance. Many of the initiatives taken were secret. Secret laws, secret courts, secret policies. Even a decade after 9/11, it wasn't possible, as an ordinary citizen, to know what the state of surveillance and civil liberties in the United States was because the rules governing society were undisclosed.[24] In fact, most of what we now know about mass surveillance in the United States we learned through the revelations of Edward Snowden, a National Security Agency (NSA) contractor turned whistleblower, in 2013.[25]

The extent of government surveillance powers in the United States after 9/11 is astonishing. The details would take a whole book,[26] but here is a taste. The NSA collected data from Microsoft, Yahoo, Google, Facebook, YouTube, Skype, and Apple, among other companies, through a programme called PRISM. That included emails, photos, video and audio chats, browsing history, and all data stored on their clouds. As if that wasn't enough, the NSA also performed upstream collection of data – that is, collecting data directly from private-sector internet infrastructure, from routers to fibre-optic cables.[27]

The NSA then used XKEYSCORE to organize all the data it had collected. XKEYSCORE was a kind of search engine that allowed analysts to type in anyone's address, telephone number, or IP address and then go through their recent online activity. Everyone's communications were in there. Analysts could also watch people live as they went online and typed, letter by letter.[28]

Most of the world's internet traffic passes through infrastructure or technologies that are under the control of the United States.[29] That means that the NSA can surveil almost every internet user around the world. If you happen to be of interest to the NSA, then surveillance can be even more intrusive, as they have programmes that can access every corner of an individual's digital life and tamper with it.[30] If it serves its purposes, the NSA can then share some of the data it has collected with other intelligence agencies in allied countries. The NSA wants to collect it all and keep a permanent record of everything.[31] The intelligence community likes to call such invasion of privacy 'bulk collection of data' to avoid its straightforward name: *mass surveillance*.

The most lamentable aspect of our loss of privacy is that it didn't even help prevent terrorism. The idea that if we have

more data on people we will be able to stop bad things like terrorism from happening is an intuitive one. The appeal is
understandable. But it's wrong. Every piece of evidence we
have suggests that mass surveillance in the United States has
been utterly unhelpful in preventing terrorism. The President's
Review Group on Intelligence and Communications Technologies, for instance, could not find a single case in which mass
collection of telephone call records had stopped an attack.[32]

In 2004, the FBI analysed the intelligence gathered from
STELLARWIND, a surveillance programme composed of warrantless bulk phone and email collection activities, to see how
many had made a 'significant contribution' to identifying terrorists, deporting suspects, or developing a relationship with an
informant about terrorists. Only 1.2 per cent of the tips from 2001
to 2004 were useful. When the FBI looked at the data from 2004
to 2006, they found that no tips had made a significant contribution.[33] The leads sent from the NSA to the FBI were too many,
and too much of a waste of time.[34] When the NSA complained to
the FBI that they were not seeing results from the information
they were passing on, an FBI official responded with the feedback
he received from field officers: 'You're sending us garbage.'[35]

Terrorism is a rare event; it is a needle in a haystack. Throwing more hay at the haystack doesn't make finding the needle
any easier – it makes it harder. By collecting much more irrelevant than relevant data, mass surveillance adds more noise
than signal.[36] Even if mass surveillance could prevent attacks,
however, we have to bear in mind that surveillance is not without harms and risks. The risk of a terrorist attack has to be

weighed against the risk of a massive misuse of data, in addition to the erosion of civil liberties. Privacy losses kill too, as we will see.

In the two decades it has been around, mass surveillance does not seem to have prevented terrorism, but it has been very effective at stripping away the right to privacy of all internet users. Surveillance has also been used for the purposes of economic and international espionage; its targets have included ally countries and aid organizations.[37] Digital mass surveillance's main contribution was giving more power to the powerful – tech companies that soon became big tech, and governments – and disempowering ordinary citizens.

There are a few lessons to be drawn from this depressing episode in history. One is that the surveillance society was born out of the collaboration between private and public institutions. The government allowed corporate data collection to thrive so it could make a copy of the data. The data economy was allowed to run wild because it proved to be a source of power for governments. In exchange, corporations assisted government surveillance. AT&T, for example, installed surveillance equipment at the request of the NSA in at least seventeen of its internet hubs in the United States. It also provided technical help to wiretap all of the internet communications at the United Nations headquarters, a customer of AT&T.*[38]

Perhaps the best example of how surveillance capitalism is a

---

* Not all companies were as happy to collaborate in mass surveillance, but they all had to comply.

private and public venture is Palantir. Named after the omniscient crystal balls in J. R. R. Tolkien's *Lord of the Rings*, Palantir is a
secretive big data analytics company. It was founded by Peter
Thiel*[39] in 2004, funded by the CIA, and designed in collaboration with intelligence agencies. It specializes in finding insights in
troves of data.

One of the problems with XKEYSCORE, the NSA tool that
worked like a search engine, was data overload. Searching, for
example, for every IP address that made a Skype call at a particular time could return too many results. Imagine searching
through your email account, except the results are pulled from
every email account in the world. Palantir helped the NSA make
XKEYSCORE more intelligible.[40]

Such private and public cooperation remains in place. Most
countries don't have the expertise to develop surveillance and
hacking tools, so they buy them from cyberweapons manufacturers.[41] Countries around the world use the tech giants for the
purposes of surveillance. Palantir, Amazon, and Microsoft, for
instance, have provided tools that aided the Trump administration in surveilling, detaining, and deporting immigrants[42] –
policies that have been exceptionally controversial on account of
the mass separation of children from their parents.[43] Shopping

---

* Peter Thiel is a billionaire entrepreneur and venture capitalist with
  notoriously illiberal views. He has written that he does not 'think
  that freedom and democracy are compatible'. Given that he con-
  siders freedom 'a precondition for the highest good', the implication
  seems to be that he no longer supports democracy.

malls in the United States have been known to capture licence plates and add those images to databases used by law enforcement agencies. Big companies sometimes also instruct police on how to access and use the data of their customers. In August 2019, representatives from AT&T, Verizon, Sprint, T-Mobile, and Google were advertised as speakers at a seminar entitled 'Wireless Carrier and Internet Provider Capabilities for Law Enforcement Investigators', which included topics like 'interpretation and usage of cellular data'.[44] Court records in the United States show that investigators can ask Google to disclose everyone who searched for a particular keyword (as opposed to asking for information on a known suspect).[45] It is probably not a coincidence that one of Amazon's new headquarters is so close to the Pentagon. Given their close ties, it doesn't make much sense to distinguish between government and corporate surveillance. We have to tackle both.

If we only protect ourselves from governments, businesses will surveil us and pass on that information to governments. In 2018, John Roberts, Chief Justice of the Supreme Court in the United States, authored a majority opinion ruling against the government obtaining location data from mobile phone towers without a warrant. He argued that, 'When the government tracks the location of a cellphone it achieves near perfect surveillance, as if it had attached an ankle monitor to the phone's user.' The ruling clawed back some privacy for Americans, almost two decades after 9/11. But in the context of the surveillance economy, it didn't make a difference, because there are many ways of obtaining data. Rather than asking mobile phone companies for their data, the Trump administration bought access to a commercial database that maps the movements of millions of mobile phones in the United States. Given that such data is for sale from data brokers, the government doesn't need a warrant to get it. In other words, by outsourcing surveillance to private companies, the government has found a way to bypass Supreme Court rulings.[46]

Yet if we only protect ourselves from corporate surveillance, governments will collect data and pass it on to businesses. The flow of information goes both ways. Data about millions of patients in the National Health Service in the United Kingdom has been sold to pharmaceutical companies, for example.[47] Big tech companies were quick to open conversations with governments across the world to tackle the coronavirus pandemic through the use of smartphone apps. The pandemic takes us to a second lesson to be learned from surveillance after 9/11: crises are dangerous for civil liberties.

During crises, decisions are taken without carefully considering pros, cons, evidence, and alternatives. Whenever there is the slightest resistance to a proposed extreme measure, an appeal to 'saving lives' silences dissenters. No one wants to get in the way of saving lives, even when there is not a shred of evidence that our initiatives will in fact prevent deaths. Government agencies and corporations engage in power grabbing. Civil liberties are sacrificed unjustifiably without proper guarantees of getting them back after the crisis. Extraordinary measures taken in the grip of panic tend to remain in place long after the emergency is over.

'Never again' was an unrealistic response to the 9/11 terrorist attacks. It was a simplistic and absurd catchphrase that warped policy discussions for more than a decade.[48] Invulnerability is unattainable. We should distrust any policy that promises zero risk. The only place where you will find zero risk is six feet underground, once you have stopped breathing. To live is to risk, and to live well is to manage risk without compromising that which makes for a good life. Terrible things like terrorist attacks and epidemics happen in the world – and they will continue to happen. To think we can prevent them if we give up our freedom and privacy is to believe in fairy tales. Such wishful thinking will only lead us to add authoritarianism to the list of catastrophes that we will have to endure. Ironically, authoritarianism is one disaster we can avert. But we have to defend our civil liberties in order to do so. That means keeping your personal data safe. Being too risk-averse can paradoxically drive us into bigger dangers down the line.

But it's hard to remember the value of privacy in the midst of an emergency. When we become afraid for our lives, the last thing in our minds is our personal data. The dangers of terrorism and epidemics are tangible in a way that privacy threats are not. A terrorist attack instantly leaves dead bodies to mourn that also serve as a warning to those left standing. An epidemic is less immediate in its effects – the 2020 coronavirus seemed to take one to two weeks to land someone in hospital – but also leaves a death trail that can understandably strike terror into the hearts of the population. Yet privacy losses can be lethal as well, both in the literal sense and the metaphorical sense of killing ways of life that we value; it's just that the casualties typically take much longer to materialize.

Data collection doesn't cut our flesh or make us bleed, it doesn't infect our lungs making it hard to breathe. But data collection is poisoning our lives, institutions, and societies. It just takes time for the consequences to unfold. Personal data is toxic, but it's a slow-acting poison.

We are more likely to protect our personal data adequately, even during a crisis, if we always keep in mind why privacy is important.

FORGETTING WHAT MATTERS AND WHY IT MATTERS

In 2010, Facebook founder Mark Zuckerberg suggested that privacy was no longer 'a social norm', that we had 'evolved' beyond it. 'People have really gotten comfortable not only sharing more information and different kinds, but more openly and with more people,' he said.[49] His statements don't seem true or sincere, given that he bought the four houses surrounding his own in order to have more privacy.[50] And let's never forget that Facebook's entire stream of income depends on exploiting our personal data. A month before Zuckerberg declared privacy dead, Facebook had controversially changed the default settings on the platform to compel users to share more information publicly.[51] It is in the interest of tech giants that we believe that privacy is outdated, that it is an anachronism. But it's not.

An anachronism is something that belongs to a different time period, typically something that has become obsolete. We often inherit objects, social norms, and laws that were made for a very different context and have become dysfunctional in the present time. Some anachronisms are plain funny. Christ Church, my previous college at the University of Oxford, used to allow only the Porter to keep a dog in college. To bypass this outdated rule, as recently as the 1990s the Dean's dog was officially considered a cat.[52] Another example: Members of Parliament in the United Kingdom are not allowed to wear a suit of armour in Parliament.[53] Not all anachronisms are funny though – some are positively pernicious.

Countries often have many thousands of laws and regulations. It's hard to keep track of them all, and laws that should no longer apply are not always repealed. Anachronistic laws are dangerous because they may be used in questionable ways. For instance, an obscure New York law from 1845 against wearing masks was used in 2011 to arrest Occupy Wall Street protesters, in contrast to other people wearing masks (think Halloween and epidemics) who do not and will not get arrested.

We have very good reason to do away with anachronistic laws and norms, as they can lead to unfairness and to the delay of progress. That is why Zuckerberg hinting that privacy had become obsolete was so significant. Since then, wanting to reassure users and keep up with competitors who are more serious about privacy, Zuckerberg has changed his tune and in 2019 claimed that 'the future is private'.[54] Just one month later, however, Facebook's lawyer argued in court that users had 'no privacy interest', because by the sheer act of using the platform they had 'negated any reasonable expectation of privacy'.[55] If Zuckerberg is right that the future is private, and Facebook's lawyer is right that users cannot hope for privacy on the platform, then the logical conclusion seems to be that the future is one in which there is privacy and Facebook doesn't exist.

Even though Zuckerberg has backtracked on his words, privacy still gets routinely blamed for being an obstacle to progress. Since 2001, privacy has been accused time and again of being a hindrance in the authorities' efforts to keep citizens safe. Privacy also gets a hard time in medical contexts. Physicians and tech companies hungry for personal data argue that privacy is a barrier to the advancement of personalized medicine and big data analytics.

During the coronavirus pandemic there was much discussion about how loosening privacy standards could help tackle the outbreak. Countries around the world used a variety of contact-tracing apps to try to identify people who might have the infection. Experts investigated to what extent the laws in their countries allowed for exceptions to data protection in the context of a pandemic. A report by the Tony Blair Institute for Global Change argued that a dramatic increase in technological surveillance was 'a price worth paying' in the fight against the coronavirus, even if 'there are no guarantees that any of these new approaches will be completely effective'.[56]

Just as dangerous as keeping obsolete norms is mistaking crucial norms for obsolete ones. Privacy has a long history. We can find evidence of some form of privacy norms in almost every society that has ever been studied.[57] To those who say that privacy is dead, ask them for their password to their email account. Or, better yet, next time they're in a public toilet, greet them from the adjoining cubicle as you take a peek over the divider. You won't be disappointed – privacy norms are in good health.

Privacy's relative success in managing to stay alive over time and through many cultures can constitute its own risk by making it easy for us to take it for granted. Privacy's benefits have been stable enough for long enough that we can forget how much it matters and why. A similar phenomenon is common in the context of public health. When we successfully prevent or rapidly contain an epidemic, for instance, we are likely to underestimate the importance of those measures when we next face the threat of an epidemic because we didn't get to experience the bad effects of what could have happened without our intervention. In the same way, we can forget the value of privacy if we haven't recently felt the consequences of its loss. It's not a coincidence that Germany is more privacy-conscious than most other countries. The memory of the Stasi, the security service of the German Democratic Republic, is still fresh.

In the offline world, there are often certain signals, usually quite palpable, that alert us when privacy norms have been broken. There are few sensations as socially uncomfortable as being scrutinized by someone when you don't want to be watched. When someone steals your diary, they leave a noticeable absence behind. You can catch someone spying on you through your window. The digital age was able to make us forget our privacy norms in large part because it was able to separate them from these tangible cues. The theft of digital data does not create any sensation, it does not leave a visible trace, there is no absence to perceive. The loss of privacy online only hurts once we have to bear the consequences – when we are denied a loan or a job or insurance, when we are humiliated or harassed, when we become victims of extortion, when money disappears from our bank account, or when our democracies are harmed.

The next two chapters are a reminder of two important privacy lessons that our parents and grandparents probably understood better than us: that the battle for your privacy is a struggle for power, and that personal data is toxic.

THREE

PRIVACY IS POWER

Imagine having a master key for your life: a key or password that grants you access to the front door of your home, your bedroom, your diary, your computer, your phone, your car, your safe deposit, your health records. Would you go around making copies of that key and giving them out to strangers? Probably not. So why are you willing to give up your personal data to pretty much anyone who asks for it?

Privacy is like the key that unlocks the aspects of yourself that are most intimate and personal, that most make you *you*. Your naked body. Your sexual history and fantasies. Your past, present, and possible future diseases. Your fears, your losses, your failures. The worst things you have ever done, said, and thought. Your inadequacies, your mistakes, your traumas. The moment in which you have felt most ashamed. That family relation you wish you didn't have. Your most drunken night.

When you give that key, your privacy, to someone who loves you, it will allow you to enjoy closeness, and they will use it to benefit you. Part of what it means to be close to someone is sharing what makes you vulnerable, giving them the power to hurt you, and trusting that person never to take advantage of the privileged position granted by intimacy. People who love you might use your date of birth to organize a surprise birthday party for you; they'll make a note of your tastes to find you the perfect gift; they'll take into account your darkest fears to keep you safe from the things that scare you.

Not everyone will use access to your privacy in your interest, however. Fraudsters might use your date of birth to impersonate you while they commit a crime; companies might use your tastes to lure you into a bad deal; enemies might use your darkest fears to threaten and blackmail you. People who do not have your best interests at heart will exploit your data to further their own agenda. And most people and companies you interact with do not have your best interests as their priority. Privacy matters because the lack of it gives others power over you.

You might think you have nothing to hide, nothing to fear. You are wrong – unless you are an exhibitionist with masochistic desires about suffering identity theft, discrimination, joblessness, public humiliation, and totalitarianism, among other misfortunes. You have plenty to hide, plenty to fear, and the fact that you don't go around publishing your passwords or giving copies of your keys to strangers attests to that.

You might think your privacy is safe because you are a nobody – nothing special, interesting or important to see here. Don't underestimate yourself. If you weren't that important, businesses and governments wouldn't be going to so much trouble to spy on you.

You have the power to lend your attention, your presence of mind. People are fighting for it.[1] Everyone in tech wants you to pay attention to their app, their platform, their ads. They want to know more about you so they can know how best to distract you, even if that means luring you away from quality time with your loved ones or basic human needs such as sleep.[2] You have money, even if it's not much. Companies want you to spend your income on them. Hackers with extortion on their minds are eager to get hold of your sensitive information or images.[3] Insurance companies want your money too, as long as you are not too much of a risk, and they need your data to assess that.[4] Maybe you are part of the workforce. Businesses want to know everything about who they are hiring, including whether you might be someone who will want to fight for your rights.[5]

You have a body. Public and private institutions want to know more about it, perhaps experiment with it, and learn more about other bodies like yours. You have an identity. Criminals want to use it to commit crimes in your name and let you pay the bill.[6] You have personal connections. You are a node in a network. You are someone's offspring, someone's neighbour, someone's teacher or lawyer or barber. Through you they can reach other people. That's why apps ask you for access to your contacts. You have a voice. All sorts of agents want to use you as their mouthpiece on social media and beyond. You have a vote. Foreign and national forces want you to vote for the candidate who will defend their interests. As you can see, you are a very important person. *You are a source of power.*

By now most people are aware that their data is worth money. But your data is not valuable only because it can be sold. Facebook does not technically sell your data, for instance.[7] Nor does Google.[8] They sell the power to influence you. They keep your data so that they can sell the power to show you ads, and the power to predict your behaviour. Google and Facebook are only technically in the business of data; they are mostly in the business of power. Even more than monetary gain, personal data bestows power on those who collect and analyse it, and that is what makes it so coveted.

POWER

The only thing more valuable than money is power. Power can get you anything and everything. If you have power, you can have not only money, but also the ability to get away with anything you want to do. If you have enough of it, it can even allow you to be above the law.

There are two aspects to power. The first is what philosopher Rainer Forst defined as 'the capacity of A to motivate B to think or do something that B would otherwise not have thought or done'.[9] Powerful people and institutions can make you think and do things. The means through which the powerful enact their influence are varied. They include motivational speeches, recommendations, ideological narratives of the world, seduction, and credible threats. In the digital age, they can include sorting algorithms, persuasive apps, personalised ads, fake news, fake groups and accounts, and repeating narratives that paint tech as the solution to our every problem, among many other means in which power is exerted. We can call this *soft power.*

Forst argues that brute force or violence is not an exercise of power, for people subjected to it do not 'do' anything; rather, something is done to them. I disagree. Brute force is clearly an instance of power. It is counterintuitive to think of someone who is subjecting us to violence as powerless. Think of an army dominating a population, or a thug strangling you. Max Weber, one of the founders of sociology, describes this second aspect of power as the ability of people and institutions to 'carry out their own will despite resistance'.[10] We can call this *hard power.*

In short, powerful people and institutions make us act and think differently from the way we would in the absence of their influence. If they fail to influence us into acting and thinking as they want us to, powerful people and institutions can exercise force upon us – they can do unto us what we will not do ourselves.

There are different types of power: economic, political, military, and so on. But power can be understood as an analogy to energy: it can transform itself from one kind into another.[11] A company with economic power can use its money to gain political power through lobbying, for instance. A politically powerful person can use his power to earn money through exchanging favours with private companies.

That tech giants like Facebook and Google are powerful is hardly news. But exploring the relationship between privacy and power can help us better understand how institutions amass, wield, and transform power in the digital age, which in turn can give us tools and ideas to resist more successfully the kind of domination that is kindled by violations of the right to privacy. To fully grasp how institutions accumulate and exercise power in the digital age, first we have to look at the relationship between power and knowledge.

POWER AND KNOWLEDGE

There is a tight connection between power and knowledge. At the very least, knowledge is an instrument of power. The English philosopher and statesman Francis Bacon understood that knowledge in itself is a form of power. More than three centuries later, the French historian of ideas Michel Foucault went even further and argued that power produces knowledge as much as the reverse.[12] There is power in knowing, and knowledge in power. Power creates knowledge and decides what gets to count as knowledge. Through collecting your data and learning about you, Google becomes empowered, and that power allows Google to decide what counts as knowledge about you through its use of your personal data. If Google classifies you as a middle-aged man without a college degree and suffering from anxiety, say, *that* gets to count as knowledge about you – even if it's completely wrong, or out of context, or outdated, or irrelevant. Through protecting our privacy, we prevent others from being empowered with knowledge about us that can be used against our interests. By having more power, we have more of a say in what counts as knowledge. We should partly get to decide what counts as knowledge about ourselves; we should have a say in what others are allowed to perceive or infer about us.

The more someone knows about us, the more they can anticipate our every move, as well as influence us. One of the most important contributions of Foucault to our understanding of power is the insight that power does not only act upon human beings – it constructs human subjects.[13] Power generates certain mentalities, it transforms sensitivities, it brings about ways of being in the world. In that vein, political and social theorist Steven Lukes argues that power can bring about a system that produces wants in people that work against their interests.[14]

People's desires can themselves be a result of power, and the more invisible the means of power, the more powerful they are. An example of power shaping preferences is when tech uses research about how dopamine works to make you addicted to an app. Dopamine is a neurotransmitter involved in motivating you into action through anticipating the way you will feel after your desires are met. Imagining how good a chocolate cake will taste motivates you to buy it and eat it. Anticipating how validated you'll feel when your friends like the way you look motivates you to take a selfie and share it online. Tech companies use tactics such as creating irregularly timed rewards (that's what makes slot machines so addictive) and using flashy colours to get you to engage as much as possible with their platform. 'Likes' and comments on your posts 'give you a little dopamine hit'.[15] Your desire to engage with a persuasive app does not arise from your deepest commitments and values. You usually don't wake up thinking 'today I want to spend three mindless hours scrolling through Facebook's infinite newsfeed'. Your desire is produced by the power of tech. In that sense, it is not entirely yours. Another example is political campaigners researching your convictions and affective and cognitive inclinations to show you an ad that will nudge you towards acting the way they want you to act.

Power derived from knowledge, and knowledge defined by power, can be all the more dominant when there is an asymmetry of knowledge between two parties. If, say, Facebook knows all there is to know about you, and you know nothing about Facebook, then Facebook will have more power over you than if both parties knew equal amounts about each other. The asymmetry becomes starker if Facebook knows everything there is about you, and you think Facebook knows nothing, or you don't know how much Facebook knows. That makes you doubly unknowing.

The power that comes about as a result of knowing personal details about someone is a very particular kind of power, although it also allows those who hold it the possibility of transforming it into economic, political, and other kinds of power.

POWER IN THE DIGITAL AGE

The power to forecast and influence derived from personal data is the quintessential kind of power in the digital age.

Governments know more about their citizens than ever before. The Stasi, for instance, only managed to have files on roughly a third of the population of East Germany, even if it aspired to have complete information on all citizens.[16] Intelligence agencies today hold much more information on all of the population. For starters, a significant proportion of people volunteer private information on social networks. As filmmaker Laura Poitras put it, 'Facebook is a gift to intelligence agencies.'[17] Among other possibilities, that kind of information gives governments the ability to anticipate protests and arrest people pre-emptively.[18] Having the power to know about organized resistance before it happens and being able to squash it in time is a tyranny's dream.

Tech companies' power is constituted, on the one hand, by having exclusive control of our data, and on the other, by the ability to anticipate our every move, which in turn gives them opportunities to influence our behaviour, and sell that influence to others – including governments.

Part of why big tech took us by surprise was because their ways escaped the radar of antitrust authorities. We are used to measuring the power of companies in economic terms by what they charge users. But big tech's power comes from what they take in personal data, not what they charge. Traditionally, the most common symptom of a company that deserved antitrust attention was its being able to raise prices without losing customers. Given that Google and Facebook provide 'free' services, the heuristic fails. Yet the traditional litmus test should be seen as an instance of a more general principle: if a company can mistreat its customers (through higher prices than what is fair, exploitative data practices, bad security, or other abusive conditions) without losing them, then there is a good chance it might be a monopoly.

Companies that earn most of their revenues through advertising have used our data as a moat – a competitive advantage that has made it impossible for alternative businesses to challenge tech titans.[19] Google's search engine, for example, is as good as it is partly because its algorithm has much more data to learn from than any of its competitors. In addition to keeping the company safe from competitors and helping it train its algorithm, that amount of data allows Google to know what keeps you up at night, what you desire the most, what you are planning to do next, what you are undecided about. The company then whispers this information to other busybodies that want to target you for ads.

Data vultures are incredibly savvy at using both aspects of power discussed here: they make us give up our data, more or less voluntarily, and they also steal it from us, even when we try to resist.

Tech's hard power

When data is snatched away from us even when we try to resist, that's tech's hard power. Like Google storing location data even when you've told it not to: an Associated Press investigation in 2018 found that Google was storing location data even when people had turned off Location History. Google's support page on that setting stated: 'You can turn off Location History at any time. With Location History off, the places you go are no longer stored.' That wasn't true. For instance, Google Maps automatically stored a snapshot of your latitude and longitude the moment you opened the app, even if your Location History was off. Similarly, some searches unrelated to where you are such as, say, 'recipe for chocolate chip cookies' saved your location to your Google account. To turn off location markers, you had to turn off an obscure setting that did not mention location called 'web and app activity', which was enabled by default, of course, and stored information from Google apps and websites to your account.[20]

Tech's hard power can sometimes be confused with soft power because it doesn't look as violent as other forms of hard power, such as tanks and other kinds of brute physical force. But people doing unto you what you have said 'no' to is hard power. It is forceful, and it is a violation of our rights.

Although tech's hard power has been there from the start, with companies taking our data without asking for it, their methods are becoming less subtle, more obviously authoritarian. China is a prime example. For years, the government of China has been designing and refining a system of social credit with the collaboration of tech companies. This system takes the idea of creditworthiness and exports it to all areas of life with the help of big data. Every piece of data on every citizen is used to rate that person on a scale of trustworthiness. 'Good' actions make you win points and 'bad' actions make you lose points. Buying nappies earns you points. Playing video games, buying alcohol, or spreading 'fake news' loses you points.

One of the marks of totalitarian societies is that power controls all aspects of life – it is in that sense that power is 'total'. In liberal democracies (at their best), you're not penalized in all spheres of life for small infractions committed in one area of life. For instance, playing loud music at home might make your neighbours hate you, and it might even earn you a visit from the police asking you to keep it down, but it will have no effect on your work life, or your financial credit score (unless you are unlucky enough to have your boss or your banker as neighbours). In China, playing loud music, jaywalking, or cheating in a video game will make you lose points in a score that is used to grant and limit opportunities in all spheres of life.

Citizens with a high score are sometimes publicly praised, and enjoy perks such as shorter waiting lists, and discounts on products and services like hotel rooms, utility bills, and loans. They can rent a car without paying a deposit, and they even get better visibility on dating sites. Citizens with a low score can be publicly shamed; they can find it hard or impossible to get a job, a loan, or to buy a property; they can be blacklisted from services like exclusive hotels, and even from travelling by plane or train.

In 2018, China shamed 169 'severely discredited' people by publishing their names and misdeeds, which included trying to take a cigarette lighter through airport security and smoking on a high-speed train.[21] According to the system's founding document, the scheme aspires to 'allow the trustworthy to roam everywhere under heaven while making it hard for the discredited to take a single step'.[22] By the end of June 2019, China had banned almost 27 million people from buying air tickets, and almost 6 million people from using the high-speed rail network.[23] During the coronavirus pandemic, Chinese surveillance went as far as forcibly installing cameras inside people's homes, or just outside their front doors, to make sure they complied with quarantine rules.[24]

When Westerners criticize the Chinese system of social disciplining, a common comeback is to argue that the West also has systems in which people get scored and suffer penalties as a result; it's just that Western systems of social credit are more opaque. Often, citizens don't even know they exist. There is some truth to this response. We are typically not fully aware of how our credit score is calculated and how it can be used, for instance. There are other kinds of scores too. Most people don't know this, but as a consumer you have secret scores that determine how long you wait when calling a business, whether you can return items at a store, and the quality of service you receive. There is no opting out of being scored as a consumer – it is something imposed on you.

Kashmir Hill, a reporter, requested her file from Sift, an American company that scores consumers. Hill's file consisted of 400 pages with years of Yelp delivery orders, messages she'd sent on Airbnb, details about her devices, and much more. While Sift provided her personal data on request, they didn't offer an explanation as to how that data was analysed to create her consumer score, nor did they tell her what impact the score had had on her life.[25]

Secret and opaque scoring systems are unacceptable. As citizens, we have a right to know the rules that govern our lives. Nonetheless, it is undeniable that the West generally enjoys more freedom and transparency than China, despite our deficits in governance (which we should fight to redress). Let our and others' shortfalls be lessons to curb hard power around us.

Another way in which tech can exercise hard power is through setting the rules we live by and not allowing us to break them. Instead of rules being something that exist primarily in writing, rules are increasingly being baked into code and enforced automatically by computers.[26] Instead of being free to drive in a bus or taxi lane and risk incurring a fine if you're found out, your future car may simply refuse to go where it's forbidden.[27]

In free societies, there is always some leeway between what laws are and what gets enforced. People are allowed to get away with minor infractions now and then because, in well-functioning societies, most people are happy to follow most rules most of the time.[28] Allowing some leeway makes room for exceptions that are hard to code into rules, like using the bus lane because you are driving someone in urgent need of care to the hospital. It also allows us to ignore outdated laws until we get to repeal them. Law enforced by tech would allow no exceptions. Experiencing tech's hard power through having every single little rule, governmental and private, enforced through code would deprive us of huge amounts of liberty.

But tech doesn't only use hard power to influence us. Tech is also brilliant at influencing us through soft power.

Tech's soft power

In some ways, soft power is more acceptable than hard power because it is less forceful. It seems less of an imposition. But soft power can be just as effective as hard power in allowing the powerful to get what they want. In addition, soft power is often manipulative – it makes us do something for the benefit of others under the pretence that it is for our own benefit. It recruits our will against ourselves. Under the influence of soft power, we engage in behaviour that undermines our best interests.

Manipulative soft power makes us complicit in our own victimization.[29] It is *your* finger scrolling down your newsfeed, making you lose precious time, and giving you a headache. But, of course, you wouldn't be hooked to the infinite scroll if platforms like Facebook weren't trying to convince you that if you don't continue swiping, you'll be missing out. When you try to resist the allure of tech, you are fighting against an army of techies trying to capture your attention against your best interests.

Loyalty cards are another example of soft power. When you are presented with a loyalty card at your local supermarket, what you are being offered is the chance to allow that company to surveil you and then influence your behaviour (through nudges such as discounts) into buying certain kinds of products that you wouldn't otherwise buy. A more subtle form of soft power is through seduction. Tech constantly seduces us into doing things we would not otherwise do, from getting lost in a rabbit hole of videos on YouTube, to playing mindless games, or checking our phones hundreds of times a day. Through enticing 'carrots', the digital age has brought about new ways of being in the world that do not always make our lives better.

In addition to technical ways of exercising soft power by design, a big part of the power of tech lies in *narratives*, in the stories that get told about our data. The data economy has succeeded in normalizing certain ways of thinking. Tech wants you to believe that, if you have done nothing wrong, you have no reason to object to their holding your data. When asked during an interview about whether users should share information with Google as if it were a 'trusted friend', the then CEO Eric Schmidt famously responded, 'If you have something that you don't want anyone to know, maybe you shouldn't be doing it in the first place.'[30] (Less well known is that he once asked Google to erase some information about him from the indexes – a request that was denied.[31] Have you noticed by now the pattern of techies wanting privacy for themselves but not for others?)

What Schmidt tried to do was shame people who are (sensibly) worried about privacy. He implied that if you're worried about privacy, you must have something to hide, and if you have something to hide, you must have done something wrong that shouldn't be allowed to remain hidden. But privacy is not about hiding serious wrongdoing.[32] It's about protecting ourselves from the possible wrongdoings of others, like criminals wanting to steal our money. It's about blinding power so that it cannot use knowledge about us to become even more powerful.

Companies also want you to think that treating your data as a commodity is *necessary* for digital tech, and that digital tech is *progress* – even when it might sometimes look worryingly similar to social and political regress.[33] More importantly, tech wants you to think that the innovations it brings into the market are *inevitable*.[34] That's what progress looks like, and progress cannot be stopped.

The narrative of progressive and inevitable technology is complacent and misleading. Power produces the knowledge, narratives, and rationality that favour and sustain it.[35] Tech tells us the stories that make it look both indispensable and good. But some of the technology developed in the last decades has not been progressive in the least – it has contributed to perpetuating sexist and racist trends.[36]

When Google Translate converts news pieces from Spanish into English, 'she' often becomes 'he'. Algorithms have been known to endorse sexist analogies – that 'man' is to 'doctor' what 'woman' is to 'nurse', and that 'man' is to 'computer programmer' what 'woman' is to 'homemaker', for example.[37] Vision algorithms label a wedding of a white bride as 'bride', 'woman', and 'wedding', while labelling the photograph of a north Indian bride as 'performance art' and 'costume'. The World Bank has warned that Silicon Valley is worsening income inequality.[38] Technological change that leads to social and political regress is not the kind of development we should be pursuing or abetting. That's *not* what progress looks like.

Furthermore, no technology is inevitable. There was nothing in history, nature, or destiny that made petrol cars unavoidable, for instance. If enormous oil reserves hadn't been discovered in the United States, and Henry Ford hadn't produced the cheap Model T, electric cars might have become much more popular than petrol ones.[39] And even though people have been talking about flying cars for decades, they may never become a reality. Whether a piece of technology gets developed and marketed depends on a host of variables related to feasibility, price, and human choice. History is littered with technological gadgets that have been jilted.

Remember Google Glass? In 2013, Google started selling a prototype of eyeglasses mounted with a tiny computer that included a video camera. It went on sale to the general public in May 2014. It was one of those products that created hype. *Time* magazine named it one of the 'Best Inventions of the Year'. Celebrities tried them on. The *New Yorker* published a long piece on it. Even *The Simpsons* made a show featuring Google Glass – though Homer called them 'Oogle Goggles'. Despite the big fuss, by January 2015 the product had been withdrawn.[40] (Instead of admitting defeat, Google said Glass was 'graduating' from one division of Google into another.)

There were at least two reasons for Google Glass's resounding failure. First, the glasses were ugly. Second, and more importantly, they were creepy. Even before anyone had seen someone wearing a pair, they were banned from bars, cinemas, casinos, and other places that didn't appreciate the prospect of having people recording one another.[41] The few people who tried them were derided as 'Glassholes' – evidence of how uncomfortable the public felt about these artefacts.

Google Glass didn't fly because people hated them. But Google is persistent. In 2017 they revived the project, this time geared towards industries like manufacturing services, to be used by workers. It wouldn't be surprising if at some point Google tries to relaunch Glass for the general public. Since 2013, they and others have consistently and consciously chipped away both at our privacy and at our resistance to privacy invasions. Facebook's Project Aria envisions replacing smartphones with smart glasses.[42] But we should remember that technology, like other kinds of social practices, depends on our cooperation for its success. We are the ultimate source of power for tech companies.

Technological development is nothing like a natural phenomenon such as gravity or evolution. *Technology doesn't happen to us – we make it happen*.[43] A Google Glass isn't going to invent and market itself. Nor does it happen by accident, like mutations.[44] It's up to us to make sure our technology fits our values and enhances our wellbeing. That technological progress is inevitable rings true because some *form* of technological change is indeed going to happen; but no *particular* technology is unavoidable, and change doesn't always amount to progress. Even after a piece of technology gets invented, we can choose how to use and regulate it.

A more truthful narrative than the one favoured by tech is that technological developments whose negative consequences outweigh their positive impact can and should be stopped. Regarding our more specific concern with privacy, a more precise narrative than the one peddled by tech is that treating data as a commodity is a way for companies to earn money, and has nothing to do with building good products; hoarding data is the means through which institutions are accumulating power; tech companies can and should do better to design the online world in a way that contributes to people's wellbeing; and we have many good reasons to object to institutions collecting and using our data the way they do, even if we have done nothing wrong.

Among those reasons is institutions not respecting our autonomy – our right to self-govern – both as individuals and as societies.[45] Here is where the harder side of power plays a role. The digital age thus far has been characterized by institutions doing whatever they want with our data, unscrupulously bypassing our consent whenever they think they can get away with it, doing unto us, in addition to making us do what they want. In the offline world, that kind of behaviour would be called theft and coercion. That it is not called what it is in the online world is yet another testament to tech's power over narratives.

If postal workers read our letters in the way that Gmail and third-party app developers have scanned our emails, they would go to jail.[46] Live geotracking, which once upon a time was only used for convicts, has now become the norm in the smartphones that everyone carries around.[47] Part of why bad tech has got away with so much is because it has found palatable ways of describing what it does. Bad tech exploits our data, hijacks our attention, and breaks our democracies – but it makes it sound so scrumptious, as if it were being done for our own benefit as part of optimizing the 'user experience'. 'Personalization' sounds like VIP treatment, until you realize it's a term used to describe techniques designed to tamper with your unique mind.

Instead of calling things by their proper names, tech has filled us with euphemisms about our digital realities.[48] As George Orwell put it, political language (and tech language is political) 'is designed to make lies sound truthful and murder respectable, and to give an appearance of solidity to pure wind'.[49] Privately owned advertising and surveillance networks are called 'communities', citizens are 'users', addiction to screens is labelled 'engagement', our most sensitive information is considered 'data exhaust' or 'digital breadcrumbs', spyware is called 'cookies', documents that describe our lack of privacy are titled 'privacy policies', and what used to be considered wiretapping is now the bedrock of the internet economy.

Tech has gone so far in seducing us through words that it has even sequestered the language of nature.[50] You used to be able to taste the sweetness of an apple, listen to birds tweet at sunrise, wade your feet in a stream, and find shapes in the clouds passing by. Now these words are mostly used to describe things that are the opposite of nature.

It is the job of thinkers and writers to challenge corporate bullshit and reclaim transparent language. Calling things by their names is a first step towards understanding our times and fighting for a better world. We have to build our own narratives, and use the words that tech tries to camouflage or avoid. We have to reclaim the power of deciding what counts as knowledge. Let's talk about what tech doesn't want us to talk about. For instance, let's talk about how bad tech is treating us, not like citizens, but like pawns in a game we didn't choose to play. Bad tech is using us much more than we are using it.

PAWNS

You are a pawn in the games data scientists are playing on their screens. Sometimes they call it 'artificial society'. Data analysts gather all the information they can on you – social media connections and posts, voting records, purchasing history, your car make and model, mortgage information, browsing history, inferences about your health, and more – and then run models to see how they can affect your behaviour.

And I do mean *you* in particular. It doesn't matter if you're a nobody – society is made up of nobodies, and that's who data-hungry institutions are interested in. When a friend was training to become a data scientist, he confided in me how his latest task had been to choose a random person halfway across the world and learn everything he could about them. He ended up studying in depth some guy in Virginia who, he learned, had diabetes and was having an affair. This random guy didn't have the slightest inkling that he was the subject of study by a data scientist. Right this moment, as you read these words, a data scientist might be studying *you*.

In a way, every one of us has countless data clones living in the computers of data scientists who are experimenting with us, with differing degrees of personalization. Like voodoo dolls, data scientists tinker with our virtual avatars. They try new things on them, and see what happens. They want to learn what makes us tick, click, buy, troll, vote. Once they learn how to manipulate our digital clones successfully, like marionettes, they try their tricks on flesh-and-blood people. That is how our virtual zombies come back to haunt us.

Wannabe tech gods would like to profile every single person in a society in order to run a simulation of that community. If you know people's personalities well enough, you can create zombie duplicates of them online and try out different interventions. You can figure out which political message works for them. Once you are confident your message will create the consequences you are after, you let it loose in the world. Through manipulating information you can sway elections, inspire an insurgency, spark a genocide, turn people against each other, distort their realities until they cannot tell what is true any more.

That is precisely what Cambridge Analytica did to aid political campaigns. First, data scientists developed an app called This Is Your Digital Life and got 270,000 Facebook users to download it. They paid each person $1 to $2 to fill in a psychometric survey that helped data analysts ascertain their personality types. The app would then download all of the users' Facebook data in order to find correlations between, for instance, personality traits and 'likes'. Facebook is an attractive object of social study because when people go around scrolling, liking, and commenting, they are not aware to what extent they are being surveilled, so they act more 'naturally'. Data scientists watching us go about (what should be) our business feel like anthropologists – except they can easily quantify every little thing.[51]

Appallingly, Cambridge Analytica's app also downloaded data from participants' Facebook friends without the latter's knowledge or consent.[52] Although data scientists didn't have the personality traits of these unknowing data subjects (because they hadn't answered a psychometric survey), they could use their Facebook 'likes' to guess them, based on studies done with the data of people who had answered the survey.

In short, Cambridge Analytica duped 270,000 people into betraying their friends – and democracies around the world – for about a dollar. Although participants willingly answered the survey, most of them probably didn't read all the terms and conditions, which in any case wouldn't have included a warning about how their data was going to be used to try to sway elections. Using people's online connections to download as much data as possible, the firm got its hands on the data of around 87 million Facebook users. They also acquired extra data from censuses and data brokers, among others. With all that information, Cambridge Analytica built a psychological warfare tool to influence politics around the world – a textbook illustration of how knowledge is power.

Cambridge Analytica dug deep into people's lives and minds. The personal data they appropriated was incredibly sensitive. It included 'private' messages, for instance. And what data scientists did with that data was close and personal. 'Data from millions of people' sounds very impersonal and abstract. But every one of those people is as real as you. Your data might be included in those millions.

Christopher Wylie is a data consultant who worked for

Cambridge Analytica before he became a whistleblower. In his book *Mindf*ck*, he describes a demonstration of the firm's tool given to Steve Bannon, who would later become chief executive of Donald Trump's campaign.[53] A data scientist asked Bannon to give him a name and a state in the United States. A simple query made a person's life appear on screen. If that had been you (and maybe it was), this group of data scientists would have been studying your life with a magnifying glass. This is what you look like, this is where you live, these are your closest friends, this is where you work, this is the car you drive. You voted for this candidate in the last election, you have this mortgage, you have this health issue, you hate your job, you worry the most about this political issue, and you're thinking about leaving your partner.

Just to make sure they were getting everything right, the data scientists then phoned the person they were stripping on their screens. Under the pretence of being researchers from the University of Cambridge conducting a survey, they asked their victim about names, views, and lifestyle. The phone calls confirmed what they already knew: they had designed a tool to get into the mind of almost anyone in the world. They had hacked the political system; they had found a way to collect and analyse so much sensitive data that they could build the most personalized political campaigns in history, with disastrous consequences.

Once Cambridge Analytica's data scientists had all the data on you that they could find, the first step was to sort you into a very specific personality category. They would score you according to the Big Five personality traits: how open to new experiences you are; whether you prefer planning over spontaneity; how much of an extrovert or an introvert you are; to what extent you are prosocial; and how prone you are to experiencing negative emotions like anger and fear.

The second step was to apply their predictive algorithms to your profile, and calculate on a scale of 0 to 100 per cent how likely you were to vote, for instance, or how likely you were to engage politically in a certain issue.

The third step was to figure out where you hung out so that they could get to you. Did you watch much TV? What about YouTube? Was there a social media platform where you spent a significant amount of time? Cambridge Analytica would then show you content specifically designed for people like you and watch to see whether it worked on you. Did you engage with their content? If you didn't, they would tweak it and try again.[54]

Data scientists at Cambridge Analytica studied people's life satisfaction. Following the logic of hackers, they looked for cracks and vulnerabilities in our minds. They identified those most susceptible to being influenced, such as people who were prone to being suspicious of others. They picked out people who exhibit what are called 'dark triad' traits – narcissism, Machiavellianism (ruthless self-interest), and psychopathy – and targeted them with the express objective of provoking anger. They inflamed trolls. They showed people blogs that made fun of people like them to make them feel attacked. They created fake pages on social media and went so far as to organize physical get-togethers which Cambridge Analytica staffers would attend undercover.[55]

At least two elements made Cambridge Analytica's digital

campaigns particularly dangerous. First, they showed dramatically different content to different people, thereby destroying our common experience. The content that was being discussed and scrutinized in the media was not what actual voters were seeing online. People subjected to tools designed to confuse can't rationally discuss a particular candidate with each other because they don't have access to the same information. It simply wasn't possible for two people to discuss calmly the lights and shades of a candidate like Hillary Clinton if one of those people thought that Clinton was connected to a child sex ring being run out of a pizzeria in Washington DC, for instance.

A second element that made Cambridge Analytica dangerous was that their campaigns did not look like campaigns. They didn't look like carefully designed propaganda. Sometimes they looked like news articles. At other times they looked like content created by ordinary users. Nobody knew that what looked like grassroots movements were in fact political campaigns orchestrated by online mercenaries – least of all the people who were being dragged into these polarized and polarizing views.

In a stunning undercover investigation, Channel 4 broadcast footage in which Mark Turnbull, then managing director of Cambridge Analytica, said: 'we just put information into the bloodstream of the internet [. . .] and then watch it grow, give it a little push every now and again . . . like a remote control. It has to happen without anyone thinking "that's propaganda", because the moment you think that [. . .], the next question is, "Who's put that out?" '[56]

Cambridge Analytica's repertoire of psychological and informational warfare undertakings was vast, and it knew no moral limits. It included targeted fake news, fear mongering (which went as far as showing extremely gory footage of actual torture and murder), impersonation, and offering highly unethical services such as 'bribery or honey trap stings, voter disengagement campaigns, obtaining information to discredit political opponents, and spreading information anonymously in political campaigns'.[57] This was the company that helped Trump win the United States presidency, and the Leave campaign in the Brexit referendum (through an associated political firm, AIQ); a company that seems to have had close ties to Russia.[58] I hope that one day, those who come after us will read about this shameful episode in history and find it hard to believe. May they feel safe in knowing that their democracies are robust and regulated enough that no one could ever try something like that and get away with it.

Cambridge Analytica has closed its doors, but many of the people who constituted it went on to found new data companies.[59] AggregateIQ, the Canadian political firm that was involved in the Brexit referendum and that, according to whistle-blower Christopher Wylie and evidence cited by the Information Commissioner's Office in the UK, had close ties to SCL (Cambridge Analytica's parent company), still stands.[60] Cambridge Analytica is just one example of something that anyone with data skills can do. In 2018, Tactical Tech, a non-governmental organization based in Berlin, had identified more than 300 organizations around the world working with political parties through data-driven campaigning.[61] Russia is notorious for its attempts to meddle in the politics of foreign countries and sow discord among fellow citizens through nefarious online interference. In 2016, two Facebook pages controlled by Russian trolls organized a protest and a counter-protest in Texas. The protest, 'Stop the Islamization of Texas', was orchestrated by a Facebook group with more than 250,000 followers called Heart of Texas and operated by a troll factory, the Internet Research Agency, in Russia. The counter-protest was likewise organized by a Russian-controlled Facebook group, United Muslims of America, which had more than 300,000 followers.[62] As this case illustrates, even though Cambridge Analytica is gone, our democracies are still in danger.

Cambridge Analytica's power stemmed from our data. Other malicious actors' power is partly derived from our data. Big tech's power comes from our data: from that fun personality questionnaire you did online to see which cartoon character you are most like (those questionnaires are designed with the sole purpose of collecting your data); from that dodgy app you downloaded that asked you for access to your contacts; from those loyalty cards you keep in your wallet.

Data scientists are playing with our lives as if they were toddler gods, grabbing at whatever they see as if it was theirs. They have moved fast and broken things, like lives, our ability to focus on one thing at a time, and democracies. As long as they have access to our data, we will continue to be their puppets. The only way to take back control of our autonomy, our ability to self-govern, is to reclaim our privacy.

PRIVACY, AUTONOMY, AND FREEDOM

Autonomy is the ability and the right to govern yourself. As an adult human being, you are capable of deciding what your values are – what is meaningful to you, what kind of life you want to lead – and of acting in accordance with those values.[63] When you make an autonomous decision, you fully own it. It is the kind of decision that expresses your deepest convictions, a choice that you can endorse upon reflection.

Individuals have a strong interest in having their autonomy respected by others. We want others to recognize and honour our ability to lead our lives as we see fit. In liberal democracies, with very few exceptions, no one, not even the government, can tell you what to think, what to say, what to do for a living, with whom to associate, or how to spend your time. You get to decide all those things and more. If you don't have autonomy, you don't have freedom, because your life is controlled by others. *Autonomy is having power over your own life.*

Autonomy is so important to individual and social wellbeing that any interference with it has to have a very solid justification – such as avoiding harm to others. To interfere with people's autonomy to increase your profits is not justifiable.

Privacy and autonomy are related because losses of privacy make it easier for others to interfere with your life. Being watched all the time interferes with the peace of mind that is needed to make autonomous decisions. When ballet legend Rudolf Nureyev decided to defect from the Soviet Union during a visit to France in 1961, he was obliged under French law to spend at least five minutes in a room by himself before signing a request for a 'sanctuary permit', thereby protecting him from the Russian officials who were trying to interfere with his choice.[64] You need time and space free from external pressures to make up your mind about what you want for yourself, and to have the freedom to carry out your desires. Think about how voting booths are designed to protect you from external pressures – if no one can see who you vote for, no one can force you to vote against your wishes.

When people know they are being watched and that whatever they do can have bad consequences for them, they tend to self-censor. When you don't search for a term for fear of how others might use that information about you, your autonomy and freedom are being limited. After Edward Snowden revealed the extent of government surveillance, searches on Wikipedia related to terrorism plummeted nearly 30 per cent, illustrating the so-called 'chilling effect' of surveillance.[65]

Others using your personal information to manipulate your desires is also a way of interfering with your autonomy, particularly when such influence is covert.[66] If you don't realize that the content you are accessing online is more a reflection of who advertising publishers or data scientists think you are, as opposed to a reflection of the outside world, it will be harder for you to act rationally and according to your own values. Autonomy requires that you be relatively well informed about the context you live in. When others manipulate your beliefs about the world and get you to believe something false that influences how you feel and live, they are interfering with your autonomy.

Tech has a track record of caring little or nothing about our autonomy. Many tech companies don't seem very interested in what *we* want. They don't make products to help us live the life we want to live, or become the people we want to be. They make products that will help them achieve *their* goals, products that squeeze as much data as possible from us for their benefit. They make apps that addict us to screens. They make us sign terms and conditions to inform us that we have few if any rights against them. Many companies would gladly reduce even more of our freedom. Such corporate disregard for autonomy is a new type of soft authoritarianism.

It is not an exaggeration to claim that Google would like to be godlike. First, it wants to be omniscient: it makes every effort to collect as much data as possible in order to know it all. Second, it wants to be omnipresent: it wants to be the platform through which we communicate with others, watch content online, search online, find our way in a city, access healthcare, and more (partly because that's the way it can collect more data). Third, it wants to be omnipotent: it would like to be able to take what it wants (i.e. our data), under its conditions, and transform the world in its favour. To this end, it spends more money than any other American company on lobbying.[67]

Eric Schmidt made it very clear that Google would like to seize control of your autonomy: 'The goal is to enable Google users to be able to ask [. . .] question[s] such as "What shall I do tomorrow?" and "What job shall I take?"'[68] In 2010, he went even further: 'I actually think most people don't want Google to answer their questions. They want Google to tell them what they should be doing next.'[69]

Google might try to convince you that its recommendations are based on your values, because it knows you so well. But remember: companies like Google have a conflict of interest, because what is best for you and for society is likely not what's best for their business. This misalignment of interests, and tech companies' appalling record of misbehaviour, give us more than enough reason not to entrust them with our autonomy. But even if companies like Google were more trustworthy, your autonomy is too important to be delegated to anyone but you.

You might think that there's nothing for you to worry about, because you can always disobey Google's advice. Even if Google Maps recommends you go one way, you can always disregard it and take a different route. We should not underestimate the influence of technology on us, however. Along with the products tech companies design, they are also designing their users by influencing our behaviour. As Winston Churchill put it, 'we shape our buildings, and afterwards, our buildings shape us'.

One of the reasons tech companies are getting so good at predicting our behaviour is because they are partly shaping it. If a company has control over a significant portion of your life through your smartphone and laptop, and it influences your life by choosing the content you can access and controlling the platforms you use to connect with others, shop, and work, then it is not hard to predict what you'll do next; after all, it is providing the options and nudging you along the way. It is creating a controlled environment for you, like *The Truman Show* (if you haven't watched the film, I recommend it).

That your autonomy is being threatened by tech should worry you. You should be master of your own life. But it is also a concern for the rest of us. Even if *you* had full autonomy, you have reason to want others in your society to have it too. The self-governing of a polity depends on individuals having autonomy – if the latter gets diminished, so does the former. For a democracy to be a democracy, its people have to have power over their own lives.

A democracy in which people are not autonomous is a sham. People whose autonomy is thin will be easily influenced into voting in a way that does not reflect their deepest convictions, but rather the ability of powerful actors to manipulate perceptions and beliefs.

We need you to protect your privacy so that we can regain our autonomy and freedom as a society. Even if you don't feel strongly about your own personal data, we – your family and friends, your fellow citizens, your fellow human beings around the world – need you to keep it safe, because privacy is a collective endeavour.

PRIVACY IS COLLECTIVE

Privacy is not only about you. That your data is 'personal' seems to imply that you are the only concerned party when it comes to sharing it. But that is a misunderstanding. Privacy is as collective as it is personal.[70] As the Cambridge Analytica disaster shows, when you expose your privacy, you put us all at risk.

Privacy resembles ecological issues and other collective action problems. No matter how hard you try to minimize your own carbon footprint, if others don't do their bit, you too will suffer the consequences of global warming. We are in this together, and we need enough people to row in the same direction to make things happen.

The collective nature of privacy has profound implications for how we think about so-called personal data. It has become fashionable to defend the view that personal data should be treated like property, that we should allow people to sell or trade their own data. Companies that allow you to be your own data broker are mushrooming. Given that capitalist societies are highly respectful of private property, it is intuitive to think that honouring personal data as property is tantamount to being respectful of privacy – but it's not.[71]

Let's imagine a friend (or maybe an enemy) gives you a home DNA kit as a 'gift'. Such kits are being sold for about £100. By mailing your saliva sample, you are giving away most or all of your rights to your genetic information.[72] That means companies like Ancestry can analyse, sell, and communicate your genetic information as they wish. Not having genetic privacy can be bad for you. For many kinds of insurance you are obliged to report genetic test results, and reporting the test can result in your being denied cover, or having to pay higher premiums. If you fail to disclose the results of your test and the insurance company learns of it (something that's quite likely, given that most DNA testing companies sell this data for a living), it can result in termination of your policies.[73]

It might be that you're willing to take these risks for yourself. Maybe you are curious about whether you have a gene that makes you sneeze in the sun (that's included in 23andMe's report),[74] or maybe you have more serious reasons for wanting to know more about your genes. But what about your family? Your parents, siblings, and children might not be happy to have their genetic privacy stripped away.[75] There is no way of knowing what the law will be like in two or three decades' time, and no way of knowing what we might be able to infer from genetic information then. Your grandchildren may be denied opportunities in the future on account of your genetic test, and they did not consent to your donating or selling their genetic data.

Even though your DNA makes you who you are, you share most of your genetic makeup with others, including very distant kin. The proportion of your genes that are specific to you is around 0.1 per cent. Think of it this way: a printed version of your genes would fill about 262,000 pages, but only between 250 and 500 of them would be unique to you.[76]

Because similarities and differences among our genes allow inferences to be made, there is no way of telling beforehand how your DNA might be used. In a best-case scenario, your DNA might help catch a dangerous criminal. That's how the Golden State Killer, a serial murderer and rapist, was caught in California in 2018. The police uploaded the DNA they had acquired at a crime scene to GEDmatch, a free online database that includes data from commercial testing. The query revealed third cousins of the criminal, whose identities then led the authorities to the suspect.[77]

You might think that's good news. No one in their right mind wants serial murderers running around free. But we should never allow a technology to run amok based on a best-case scenario. Technologies can be used in many ways, and the best practices are rarely the only ones employed. Genetic databases can be used to identify political dissenters, whistleblowers, and protesters in authoritarian countries. Even democratic countries can use commercial databases to infer the nationality of migrants and deport them.[78]

Combining an anonymous DNA sample with some other information – for instance, someone's approximate age – can be enough to narrow that person's identity to fewer than twenty people, if you start with a database of 1.3 million people. In 2018, such a search could allow the identification of 60 per cent of white Americans – even if they had never provided their own DNA to an ancestry database.[79] The more people continue to give away their genetic information, the more feasible it will be to identify anyone in the world. That is, if everything works the way it's supposed to – except sometimes it doesn't.

Genetic testing can have a high rate of false positives. It's one thing to use DNA as supporting evidence against someone who is considered a suspect based on other evidence. But to go on genetic fishing expeditions in search of a suspect is dangerous. DNA evidence sounds irrefutable. It's intuitive to think that if someone's DNA is found at a crime scene, that person has to be the criminal. But it's not that simple. There are many ways in which DNA can end up in an investigation. In the search for a criminal nicknamed the Phantom of Heilbronn, someone's DNA had been found at more than forty crime scenes in Europe. The DNA in question turned out to be from a factory worker who made the testing swabs used by police. Genetic contamination is easy. Other times, DNA tests get accidentally swapped with someone else's. Most times, genetic data is hard to read. Looking for similarities between two different genetic samples involves subjective interpretation. Mistakes are rife.[80]

You can be completely innocent of a crime and yet be turned into a suspect through a relative having done a DNA test. That's how Michael Usry became a murder suspect.[81] His father had donated his DNA for a genealogy project. Although his father's DNA was similar to the one found at the crime scene, fortunately Usry's wasn't. After a thirty-three-day wait which must've felt like much longer, Usry was cleared. Not everyone is so lucky. There are plenty of cases of people who have been wrongfully convicted on the basis of a DNA test. According to the United States' National Registry of Exonerations, false or misleading forensic evidence was a contributing factor in 24 per cent of all confirmed wrongful convictions in the country.[82] And those are only the cases we know about.

Just as we are connected to each other through our genetic makeup, so we are tied to each other in innumerable and invisible ways that make us vulnerable to each other's privacy slips. If you expose information about where you live, you are exposing your housemates and neighbours. If you give a company access to your phone, you are exposing your contacts. If you divulge information about your psychology, you are exposing other people who share those psychological traits. You and I may never have met, and we may never meet in the future, but if we share enough psychological traits and you give your data away to the likes of Cambridge Analytica, you give away part of my privacy too. Since we are intertwined in ways that make us vulnerable to each other, we are partly responsible for each other's privacy.

Our interdependence in matters of privacy implies that no individual has the moral authority to sell their data. We don't own personal data like we own property because our personal data contains the personal data of others. Your personal data is not only yours.

Privacy is collective in at least two ways. It's not only that your privacy slips can facilitate violations of the right to privacy of other people. It's also that the consequences of losses of privacy are experienced collectively. A culture of exposure damages society. It hurts the social fabric, threatens national security (more on this later), allows for discrimination, and endangers democracy.

Living in a culture in which anything you do or say might be broadcast to millions of others puts a considerable amount of pressure on people. Feeling that we can never make a mistake in public when our private spaces have shrunk puts a heavy weight on our shoulders. Almost everything you do is potentially public. Human beings are simply not the kind of creatures who can thrive in fishbowls. When we trust that others will not pass on what we say, we are more likely to be sincere, bold, and innovative.

There is no intimacy without privacy. Relationships that cannot count on the shield of confidentiality – either because we distrust others or because we distrust the technologies we use to communicate and interact with others – are bound to be more shallow. A culture of privacy is necessary to enjoy intimate conversations with others, have frank debates within closed settings such as homes and classrooms, and establish the bonds upon which well-functioning liberal societies are based. To be in a world in which data is constantly weaponized is to feel perpetually threatened and distrustful of others. Such fear breeds conformity and silence.

The 'spiral of silence' is the tendency for people not to voice their opinions in public when they believe their views are not widely shared. Research suggests that both social media and surveillance more generally lead to an increase in the spiral of silence.[83] The fear of suffering social isolation and other negative consequences pushes people to conform. When I teach a class or give a talk in a context in which everything is being recorded (or worse, live-streamed online), I often notice how I hold back in some of what I say, and my students or the audience ask less controversial questions. I am told that since trials began to be recorded in Spain, there is less camaraderie in the courtroom, and more silence.[84]

When surveillance is everywhere, it becomes safer to keep quiet, or to echo the opinions that others accept. But society progresses through listening to the arguments of those who are critical, those who rebel against the status quo.

A lack of privacy also inflicts damage on society when personal data is used for the purposes of personalized propaganda and fake news. When malicious actors disseminate personalized fake news, sometimes they have a very concrete objective in mind, like helping a particular candidate win an electoral race. But often their ultimate goal is simply to sow discord in a society. To divide and conquer is a very old political strategy that is being revamped through social media. We get divided along the lines of our personal data, and we get conquered through personalized propaganda.

Everyone is vulnerable to manipulation because no one has unmediated access to information. You can't be a first-hand witness to everything relevant that happens in your country and around the world. You learn about candidates and political events mostly through your screens. But very often you don't choose your sources. You don't go looking for them – they come looking for you. They pop up on your Twitter or Facebook feeds. And while they might seem to appear as if by magic or coincidence, companies like Facebook are carefully curating that content for you. They are selling your attention to unknown actors who want to influence you.

If you and I receive contradictory information about a candidate, and neither of us can see the information that the other has been exposed to, it's likely that when we talk about that candidate we'll think of each other as stupid, insane, or both, rather than realize that we are experiencing reality through very different filters that have been put in place specifically for us by people who want us to hate each other. When we can't see a common reality, society gets polarized, and the bad guys win. Polarized societies are more fragile. Cooperation becomes difficult, and solving problems that require collective action becomes impossible. When each of us is trapped in an echo chamber, or an information ghetto, there is no way to interact constructively.

Another way in which malicious actors disseminate discord online is through cultivating negative emotions in the population. The more we are afraid and angry, the more distrustful we will be of each other, the less rational our decisions will be, and the worse our societies will work.

The power that privacy grants us collectively as citizens is necessary for democracy – for us to vote according to our beliefs and without undue pressure, for us to protest anonymously without fear of repercussions, to have freedom to associate, speak our minds, read what we are curious about. If we are going to live in a democracy, the bulk of power needs to be with the people. And whoever has the data has the power. If most of the power lies with companies, we will have a plutocracy, a society ruled by the wealthy. If most of the power lies with the state, we will have some kind of authoritarianism. For governments' power to be legitimate, it has to come from people's consent – not from our data. Liberal democracy is not a given. It is something we have to fight for every day. And if we stop building the conditions in which it thrives, liberal democracy will be no more. Privacy is important because it gives power to the people. Privacy is a public good, and defending it is our civic duty.[85]

WHY LIBERAL DEMOCRACY?

Democracy refers to a system of government in which the sovereign power is vested in the people. Democracy aspires to have social equals govern themselves and achieve a relatively just social order without dictators or autocratic rulers.[86] Perhaps a few decades ago it would've been enough to argue that we need privacy in order to sustain liberal democracies. But these days democracy is not at the height of its popularity. Only a third of Americans under thirty-five say that it is vital to live in a democracy, and the share who would welcome military government increased from 7 per cent in 1995 to 18 per cent in 2017.[87] Around the world, civil liberties and political rights have declined in the past twelve years – in 2017, only thirty-five countries had improved, while seventy-one had lost ground.[88] The Economist Intelligence Unit described 2019 as 'a year of democratic setbacks', with the average global score for democracy at its lowest since 2006 (when the Democracy Index was first produced).[89] The attacks on democracy have only accelerated during the coronavirus pandemic. According to Freedom House, a think-tank in Washington DC, democracy and respect for human rights have deteriorated in 80 countries since the outbreak of the coronavirus.[90]

It is therefore necessary to make a case for why you should still fight for liberal democracy. Even if your current President or Prime Minister is a dimwit. Even if you think that the current government or past governments (or both) have ruined

your country. Even if you feel excluded from the political process. Even if you feel unrepresented by your local politicians.
Even if you suspect your society has been hacked. Even if you
distrust your fellow citizens – *especially* if you distrust your fellow citizens. Even if you have been let down by democracy, you
should work towards improving it, as opposed to getting rid of
it, because it is the system that most adequately protects the
fundamental rights of everyone, including you.

'No one pretends that democracy is perfect or all-wise,' Winston Churchill famously said in 1947. 'Indeed, it has been said
that democracy is the worst form of government – except all
those other forms that have been tried from time to time.'[91]

Democracy is not a great system. At its best it is messy, pain-
fully slow, and resistant to change. It is so patchy it looks like a
quilt made by a five-year-old. It requires compromises, such
that most of the time, no one gets exactly what they want, and
everyone ends up somewhat unsatisfied. At its worst, the system
is co-opted by a handful of rich people who write the rules of
society to benefit themselves at the expense of everyone else.

We can agree that democracy is no heaven on earth. But it
has advantages no other political system has. Democracy
forces politicians to take into account the interests and opinions of most people in society. Politicians depend on our
support to stay in power, which compels them to try to keep a
majority of the population reasonably happy. That democracy
involves many more people than other forms of government
gives it a better chance of making good decisions, since it can
take advantage of many sources of information and points of

view.[92] Democracies tend to be more prosperous. They also tend to be more peaceful, both within their borders and with respect to other countries (an idea expressed by the democratic peace theory, a tradition going back to Immanuel Kant).[93] The philosopher Karl Popper reminded us that democracies are the best way of getting rid of bad governments without bloodshed, and of implementing reform without violence.[94]

Yet many of the evils that can be found in authoritarian societies can be found in democracies. If you look for instances of abuses of power and unfairness you are likely to find them. But how prevalent these instances are is what makes all the difference. A difference in degree becomes a difference in kind. George Orwell argued that the best asset of democracy is 'the comparative feeling of security' that its citizens can enjoy. Talking about politics with your friends without fearing for yourself. Being confident that no one will punish you unless you have broken the law, and knowing that 'the law is above the State'.[95] That I am able to write this book, challenging the most powerful actors in our society, without fear, and that you are able to read it is evidence that we live in free societies. We should not take it for granted.

For *your* rights in particular to be guaranteed, democracy has to be liberal. Otherwise we risk what John Stuart Mill called 'the tyranny of the majority'. A majority can be as oppressive towards a minority as an autocrat. Liberalism strives to allow as much freedom to citizens as possible while making sure the rights of all are respected. Liberalism enforces just the necessary limits so that each of us can pursue our ideal of the good life without interfering with one another. If you are an ordinary citizen, living in a liberal democracy is your best chance of having the most autonomy. Liberal democracies allow us to self-govern, as individuals and as societies.

When liberalism is neglected, democracies can be destroyed through a dismantling of the system from the inside. Democracies don't always die with a violent splash – they can also die at the hands of elected leaders. Hitler in Germany and Chávez in Venezuela are two notorious examples.[96] The British philosopher Jonathan Wolff has argued that the first step of the fascist undoing of democracy is to prioritize the will of the majority over minority rights. The second step, he argues, is to question the means through which the will of the majority is expressed, thereby undermining voting.[97] (In the digital age, we should look out for claims by tech that your gadgets can interpret your will and vote for you. AI scholar César Hidalgo, for example, has argued that in the future we should have digital avatars that vote on our behalf.[98] Bad idea.)

Liberal democracy limits the rule of the majority to make sure the rights of the minority are protected. In a liberal democracy, you cannot go to prison if you have not broken the law, even if the majority of your society would vote for your rights to be violated. That's what the rule of law is there for.

PRIVACY IS THE BLINDFOLD OF JUSTICE

One of the greatest virtues of liberal democracy is its emphasis on equality and justice. No one is above the law, everyone has the same rights, everyone of age gets a vote, and everyone gets the opportunity to participate in democracy in more active ways – even the people who end up on the losing side of a vote. One of the greatest vices of the data economy is how it's undermining equality in various ways. The very essence of the personal data economy is that we are all treated differently, according to our data. It is because we are treated differently that algorithms end up being sexist and racist, as we've seen. It is because we are treated differently on account of our data that different people get to pay different prices for the same product without knowing they might be paying more than others. It is because we are treated differently that we get to see different content, which further amplifies our differences – a vicious cycle of otherness and inequality. No matter who you are, you should have the same access to information and opportunities. Personifications of justice are often depicted wearing a blindfold, symbolizing justice's impartiality. Privacy is what can blind the system to ensure that we are treated equally and impartially. Privacy is justice's blindfold.

REDRESSING POWER ASYMMETRIES

Big tech and political puppet masters have been very successful at manipulating us because we have been suffering from an asymmetry of knowledge that has led to an asymmetry of power. Until recently, we have known very little about how big tech and political propaganda work in the digital realm. Their tactics were invisible to us. Meanwhile, they know almost everything about us. We have to work to tip the balance back in our favour. We have to know more about them, and make sure they know less about us. Reading this book is a step in the right direction; it will help you to inform yourself about the power of big tech and governments. Better safeguarding of your privacy is the next step. If you keep your data safe, they will know less about you as a person and about us as a citizenry.

There are a few guardians of truth, justice, and impartiality whose independence has to be defended for the health of liberal democracies: the press, the courts, and academia. An important part of redressing power asymmetries in the digital age is supporting them. As an academic, I worry that more and more research (including research in ethics) is being funded by big tech. If big tech wants to fund research, let them fund it through intermediaries that allow space between researchers and the source of the funding. Intermediaries can be governments, independent foundations, or universities, as long as the funds are donated in a way that leaves absolutely no strings attached. If research funds can disappear when academics defend a controversial point of view, academic freedom will be compromised, and society will be the worse for it, as academics won't be able to research what they think is most important and disseminate the results of their research. Already I can see researchers avoiding controversial points of view and choosing topics that big tech will look at with benevolent eyes. If you hope to be funded by Google, do you think you're going to do research on the morally problematic aspects of ads? Just as we should be extra critical of medical research funded by big pharma and nutrition research funded by food companies, we should be wary of research funded by big tech.

During the past few years, independent journalism has been one of surveillance society's fiercest opponents – along with whistleblowers. Edward Snowden blew the whistle on mass surveillance and we learned about it thanks to Laura Poitras, Glenn Greenwald, Ewen MacAskill and the *Guardian*, then led by editor Alan Rusbridger. Carole Cadwalladr, from the *Observer*, unveiled the workings of Cambridge Analytica and gave a voice to whistleblower Christopher Wylie.

All of these people have had to withstand enormous pressures to inform us. Snowden had to seek asylum in Moscow and may never be able to go back to the United States. Greenwald's partner was detained and questioned for nine hours at Heathrow airport under the Terrorism Act, his computer seized. Laura Poitras was continually detained at airports and questioned. The *Guardian*, under threat of injunction, and under the watchful gaze of government officials, was forced to destroy the hard drives that contained Snowden's leaked documents. At the time of writing, Carole Cadwalladr is facing a libel claim from millionaire Arron Banks, who was involved in the Brexit campaign. If it weren't for brave journalists, we wouldn't be aware of the rules that we live by. Read and support good journalism. It's part of how citizens empower themselves against corporate and governmental power.

Fake news and propaganda have something in common with magic tricks. Magic tricks catch our attention and inspire awe in us, even when we know they are illusions. Former magician turned psychology professor Gustav Kuhn has found that illusions can be so compelling, even when at some level we know we are being fooled, that a good trick will still make us think something paranormal might be going on. The spell is only broken once we are told how the trick is done.[99] In the same way, understanding how personalized content gets designed and for what purposes might take away some of its power – it might break the spell.

RESISTING POWER

As these instances of brilliant and brave journalism show, it's not all bad news. Power can be resisted and challenged. You have power too, and collectively we have even more power. Institutions in the digital age have hoarded too much power, but we can reclaim the data that sustains it, and we can limit their collecting new data. The power of big tech looks and feels very solid. But tech's house of cards can be disrupted. They are nothing without our data. A small piece of regulation, a bit of resistance from citizens, a few businesses starting to offer privacy as a competitive advantage, and it can all go up in smoke.

No one is more conscious of their vulnerability than tech companies themselves. That is why they are trying to convince us that they do care about privacy after all (despite what their lawyers say in court). That is why they spend millions on lobbying.[100] If they were so certain about the value of their products for the good of users and society, they would not need to lobby so hard. Tech companies have abused their power and it is time to resist them.

In the digital age, the resistance that is inspired by the abuse of power has been dubbed a techlash.[101] Abuses of power remind us that power needs to be curtailed for it to be a positive influence in society. Even if you happen to be a tech enthusiast, even if you think that there is nothing wrong with what tech companies and governments are doing with our data, you should still want power to be limited, because you never know who will be next in power. Your next Prime Minister might be more authoritarian than the current one; the next CEOs of the next big tech companies might not be as benevolent as their predecessors.

Do not give in to the data economy without some resistance. Don't make the mistake of thinking you are safe from privacy harms, maybe because you are young, male, white, heterosexual, and healthy. You might think that your data can only work for you, and never against you, if you've been lucky so far. But you might not be as healthy as you think you are, and you will not be young for ever. The democracy you are taking for granted might morph into an authoritarian regime that may not favour the likes of you. If we give away all of our power to surveillance capitalism because we think our current leaders are benevolent, we won't be able to reclaim that power when things turn sour, either because we get new leaders or because our current leaders disappoint us. Also, remember the old adage of nineteenth-century British politician John

Dalberg-Acton: 'Power tends to corrupt, and absolute power corrupts absolutely'. It's not prudent to let tech or governments hold too much power over us.

But before we go into the details of *how* to take back control of your personal data – and with it your autonomy and our democracies – there is one more reason *why* we should resist the data economy. In addition to creating power imbalances, the surveillance economy is dangerous because it trades in personal data, and personal data is a toxic substance.

FOUR

TOXIC DATA

In many ways, asbestos is a wonderful material. It is a mineral that can be cheaply mined and is unusually durable and fire resistant. Unfortunately, in addition to being very practical, asbestos is also deadly. It causes cancer and other serious lung conditions, and there is no safe threshold for exposure.[1] Personal data is the asbestos of the tech society. Like asbestos, personal data can be mined cheaply. Much of it is the by-product of people interacting with tech. Like asbestos, personal data is useful. It can be sold, exchanged for privileges, and it can help predict the future. And like asbestos, personal data is toxic. It can poison individual lives, institutions, and societies.

Security expert Bruce Schneier argues that data is a toxic asset.[2] Every day of every week hackers break into networks and steal data about people. Sometimes they use that data to commit fraud. Other times they use it for shaming, extortion, or coercion. Collecting and storing personal data constitutes a ticking bomb, a disaster waiting to happen. In cyberspace, attackers tend to have an advantage over defenders. While the attacker can choose the moment and method of attack, defenders have to protect themselves against every type of attack at all times. The upshot is that attackers are very likely to get access to personal data if they are set on it.

Personal data is dangerous because it is sensitive, highly susceptible to misuse, hard to keep safe, and desired by many – from criminals to insurance companies and intelligence agencies. The longer our data is stored, and the more it is analysed, the more likely it is that it will end up being used against us. Data is vulnerable, which in turn makes data subjects and anyone who stores it vulnerable too.

POISONED LIVES

If your personal data ends up in the wrong hands, it can ruin your life. There is no way to foresee disaster, and once it happens, it is too late – data cannot be recalled.

On 18 August 2015, more than 30 million people woke up to find some of their most personal data published online. Hackers had released the entire customer database of Ashley Madison, a dating site that helped married people have affairs. Users (including people who had cancelled their membership) could be identified by their names, addresses, preferences, postcodes, and credit card numbers. The hackers wanted to teach cheaters a lesson. 'Make amends', they wrote.[3]

It is hard to get an accurate sense of the suffering and destruction left in the wake of this data leak. Millions endured sleeplessness and anxiety. Some lost their jobs. Some were blackmailed by criminals who threatened to tell their spouses about their use of the website if they didn't pay up. In one case, the extortion letter contained the following threat in exchange for money: 'If you don't comply with my demand I am not just going to humiliate you, I am going to humiliate those close to you as well.'[4] Even if the victim paid for the promise of silence, they could not be sure that the same criminal, or someone else, might not expose them anyway. In Alabama, a newspaper printed all the names of the people in the database from the region. A hacker opened a Twitter account and a website to publish the most salacious details they could find in the data leak, just for the fun of it. Marriages and families were broken, and there were tragic suicides.[5]

You might think that users of Ashley Madison deserved what they got because they were cheaters. Questionable. It is wrong to think that whoever is guilty of something is fair game for social punishment. We are all guilty of *something*, at least in the eyes of some. Nonetheless, everyone has a right to privacy, and hackers do not have the moral legitimacy in our society to judge and punish people. Moreover, online shaming and the loss of a job are not appropriate punishments for being unfaithful. Also, bear in mind that some users of the site had complicated reasons for being on it, and were not as guilty as they might seem at first glance. Some were there with the knowledge and consent of their spouse. Others were there because their spouses refused to sleep with them. Others had signed up to the site in a moment of weakness but had never done anything with their membership – their signing up was only a reminder that they could look for a connection outside their marriage, should they want to. Even if you think that Ashley Madison users had it coming, their innocent spouses and children certainly did not deserve the public humiliation they had to endure by association.

Reading about this data disaster, you might sigh in relief and think that you are safe because you have never lied to your family. (Your family might be lying to you, though.) But you don't have to have dark secrets for your personal data to poison your life. It can be something as banal as your passport or identity card, or your name, address, and bank details.

Two men woke up Ramona María Faghiura in the middle of the night in January 2015. They showed her an arrest warrant and took her into custody. She assured the policemen that she had done nothing wrong, to no avail. She tried to explain that she had been the victim of identity theft, that the person they were looking for was not her. It was no use. She texted her husband as she sat crying in the back of the police van: 'They are arresting me. Bring the folder.' The folder contains a summary of her nightmare: court papers, subpoenas, bail bonds, and her complaints before judges and police, reporting time and again that someone had used her identity to commit fraud in a dozen Spanish cities.

Ramona María Faghiura did nothing wrong. Yet she has spent years coming in and out of police stations and courtrooms, paying out thousands of euros to lawyers in the hope they'll prove her innocence. She has been diagnosed with anxiety, and has to take medication for it. 'My life has been ruined,' she lamented.[6]

Cases of identity theft have become a relatively common experience in the digital age, with credit card fraud being its most frequent form. Every year we add more information online, create more public databases that criminals can use to infer personal data, yet our security protocols are not improving. We shouldn't be surprised that data-related wrongs are becoming commonplace. In a recent survey my colleague Siân Brooke and I carried out, a staggering 92 per cent of people reported having experienced some kind of privacy breach online, ranging from identity theft to public humiliation and being the target of spyware.[7]

Another increasingly widespread data-related crime is extortion. In 2017, a criminal group gained access to data from a Lithuanian cosmetic surgery clinic and blackmailed patients, who came from sixty countries around the world, asking for a bitcoin ransom. Hackers ended up publishing more than 25,000 private photos, including nude ones, and personal data including passport scans and national insurance numbers.[8]

Patients of a large psychotherapy clinic in Finland have been recently blackmailed after a hacker stole their data. About 300 records were published on the dark web.[9]

In 2019, a Spanish woman, mother of two small children, died by suicide after a sexual video of her was shared among her work colleagues through a WhatsApp group chat.[10]

False rumours spread over WhatsApp about child abduction have led to innocent people being beaten and lynched in India.[11]

A Japanese man accused of stalking and sexually assaulting a woman told police he was able to find her through the reflection in her eyes in a photograph shared on social media; he used Google Street View to find a matching bus stop.[12]

In the United States, the Detroit Police Department wrongfully arrested a man solely based on a flawed match from a facial recognition algorithm.[13]

Researchers believe that a hacked phone might have led killers to journalist Jamal Khashoggi, who was murdered in Turkey in 2018.[14]

These are only some examples of the many ways in which the lives of countless people get poisoned every year through the misuse of personal data. Stories of revenge porn, online shaming, exposure, and other violations of the right to privacy abound. It is not only data subjects who suffer the consequences. Data disasters can damage governments and companies too.

POISONED INSTITUTIONS

Data's vulnerability spills into the institutions that store and analyse it. Any data point can trigger a disaster that may diminish a company's profits, damage its image, reduce its market share, hurt its stock price, and potentially result in expensive lawsuits or even criminal charges. Institutions that hoard more data than is necessary are generating their own risk.

It is true that not all companies are ruined by data disasters. Some companies have been lucky. Ashley Madison, for instance, is doing better than ever. Facebook has survived its countless data blunders. Its image has taken a knock, however. It might feel like it is still socially or professionally necessary to be on Facebook, but it is no longer *cool*. In itself, that might spell the end of Facebook in the long run. The company might run into trouble as soon as a competitor arises that can offer a meaningful alternative. In our recent survey, people judged Facebook to be the least trustworthy company of all the tech giants. Participants gave Facebook a mean score of 2.75 on a scale from 0 ('I don't trust them at all') to 10 ('I trust them completely').[15]

No one would be surprised if Facebook ended up losing its position of power as a result of its disrespect for privacy. That prospect remains to be seen, however. Facebook is still standing, even though working there has become a source of shame for some of its employees and ex-employees, rather than pride. When I started working in digital topics, people working for the social media giant would tend to brag about their jobs. These days it is not uncommon for people to keep quiet about their ties to the blue thumbs-up.[16]

That some companies manage to survive a data disaster does not mean that all companies do. Managing sensitive data is like managing any other toxic substance. When things go wrong, it can mean the death of a company. Take Cambridge Analytica. Two months after it was revealed that the firm tried to influence political campaigns across the world through the use of personal data to profile and microtarget voters, the firm filed for insolvency proceedings and had to close down operations. Similarly, Google shut down its social network Google+ after it was disclosed that software design flaws allowed outside developers access to users' personal data.

Even if institutions manage to get through a scandal, surviving data poisoning can be expensive. So far, Facebook has been fined $5 billion in the United States for its many privacy transgressions,[17] and £500,000 in the United Kingdom for the Cambridge Analytica scandal – but that was before the days of the European General Data Protection Regulation (GDPR).[18] The new regulation allows fines of up to 4 per cent of revenues or €20 million, whichever is the bigger number, and Facebook is currently under multiple GDPR investigations. In 2019, the Information Commissioner's Office in the UK announced its intention to fine British Airways £183 million under the GDPR for a breach of its security systems that affected 500,000 customers.[19] As long as institutions don't change their ways, we will be seeing more and possibly steeper fines. Good regulation makes sure that the interests of customers and businesses are aligned. If users are hurt by data negligence, companies must hurt too.

Sometimes a privacy disaster will not end up with a fine, but can nonetheless severely damage an institution. Such was the case with the 2015 data breach of the United States Office of Personnel Management. Hackers stole about 21 million records from the government, including background investigations of current, former, and prospective federal government employees. Among the sensitive data lost were names, addresses, dates of birth, job and pay histories, lie detector results, accounts of risky sexual behaviour, and more than 5 million sets of fingerprints. These stolen records could potentially be used to unmask undercover investigators, for instance.[20] Data breaches such as this one not only constitute a significant loss of face – they can also compromise the security of a whole country.

POISONED SOCIETIES

There are four main ways in which the mismanagement of personal data can poison societies. Personal data can jeopardize national security, it can be used to corrupt democracy, it can threaten liberal societies by promoting a culture of exposure and vigilantism, and it can endanger the safety of individuals.

Threats to national security

Equifax is one of the largest data brokers and consumer credit reporting agencies in the world. In September 2017, it announced a cybersecurity breach in which criminals accessed the personal data of about 147 million American citizens. The data accessed included names, social security numbers, birth dates, addresses, and driver's licence numbers. It is one of the biggest data breaches in history. So far, so concerning. In February 2020, the story took an even darker turn when the United States Department of Justice indicted four Chinese military people on nine charges related to the breach (which China has so far denied).

Why would the Chinese military want all that personal data? One possibility is that they wanted to identify potential targets to recruit them as spies. The more information you have about people, the better chance you have of getting what you want from them. If you realize they have debts, you can offer money. If you discover a secret, you can blackmail them. If you research their psychology, you can anticipate what makes them tick.

China has been known to use social media like LinkedIn to recruit spies. LinkedIn has 645 million users seeking employment opportunities who are open to being contacted by strangers. People who have worked for the government sometimes advertise their security clearance to improve their chances of getting hired. All this information and access is valuable to Chinese spies. It is much less risky and much more cost-effective to approach people online than in person. It is thought that Chinese spies have tried to come into contact with thousands of German and French citizens on social media.[21]

A second reason why foreign countries might crave personal data is to train their algorithms. China has troves of data on its citizens, but not enough on other people around the world. Algorithms trained on data from Chinese people might not work on Westerners. A third reason is to use that data to design targeted disinformation campaigns, much like Cambridge Analytica did. Finally, data is just another commodity that can be sold to other governments.[22] Perhaps Russia or North Korea are just as interested as China in knowing more about Americans.

The Equifax attack was carried out by professionals. The hackers stole the information in small instalments in order to avoid detection, and routed internet traffic through thirty-four servers in over a dozen countries to cover their footprints. But Equifax was apparently reckless. A class action lawsuit alleged that sensitive information was stored in plain text (unencrypted), was easily accessible, and that in at least one case, the company had used a weak password ('admin') to protect a portal. Just as importantly, they failed to update their Apache Struts software.[23] Apache had disclosed a vulnerability in its software and had offered its users a patch, but Equifax had not installed it.[24]

In the case of Equifax, data was stolen. But given the proliferation of data brokers, data can be legally bought and used for equally nefarious purposes. When the *New York Times* received location data held by a data broker from sources who 'had grown alarmed about how it might be abused', they investigated just how dangerous such data might be.[25] The data included information about 12 million phones in the United States. Sensitive areas visited by some of the people tracked included psychological facilities, methadone clinics, queer spaces, churches, mosques, and abortion clinics. Someone working at the Pentagon visited a mental health and substance abuse facility more than once. Reporters were able to identify and follow military officials with security clearances, and law enforcement officers. Using base locations as a guide, they could infer the job title of a commander in the U.S. Air Force Reserve. Even more alarmingly, it took reporters *minutes* to deanonymize location data and track President Trump himself (through the smartphone of a Secret Service agent).[26] If any of these people has something to hide (and everyone does), such easy access to them could make them targets of blackmail. In worst-case scenarios, location data could facilitate kidnap and murder. If the people in charge of security in our countries are that easy to find, follow, and potentially compromise, then we are all at risk, and foreign actors are all too aware of how personal data makes countries vulnerable.

National security concerns were the reason behind the United States pressuring TikTok, a Chinese social network, to sell its business to an American company in 2020.[27] Before that, the United States forced the gaming giant Beijing Kunlun to sell its stake in Grindr to an American company. Grindr is a dating app geared towards gay, bi, and trans people. It holds incredibly sensitive data, including sexy conversations, nude photos and videos, real-time location, email addresses, and HIV status. If we lived in a world in which privacy was taken seriously, such an app would be required to have watertight cybersecurity and privacy features. You won't be surprised to know it doesn't. In 2018, a Norwegian research organization found that Grindr sent personal data, including HIV status, to third parties that help improve apps. According to the report, much of that personal data, such as location data, was sent unencrypted to a number of advertising companies.[28]

The United States didn't reveal the details about their concerns regarding Grindr – 'doing so could potentially reveal classified conclusions by US agencies', said a source – but given the context, these are not hard to imagine. After all, Kunlun had given Beijing-based engineers access to the personal information of millions of Americans, including private messages.[29] It is likely that some members of the US military and intelligence agencies are using the app, and China could use their data to blackmail them, or to infer American troop movements.[30] It wouldn't be the first time that an app revealed troop movements.

Like most other people, those who work for highly secretive government projects jog close to where they work and live. When military personnel in the United States shared their running routes with the fitness company Strava, it didn't occur to them that they were broadcasting the location of what were supposed to be secret army bases. Strava published the running routes of all its users in a heatmap on their website on which you could zoom in and explore the most and least common paths trodden. Analysts pointed out that, not only could secret army bases be inferred by their being framed by running routes in areas with otherwise scarce activity density, but it was also possible to identify Strava users by name using other public databases. Thanks to the heatmap, one could identify and follow military individuals of interest.[31]

In this case, the data that created a national security threat was not even stolen or bought – it was public and easily accessible. After this incident, Strava made its feature to opt out of heatmaps more visible and simpler.[32] Too little too late. People should have to *opt in* for data collection. The generalized disrespect for privacy means that the privacy of military personnel and government officials is in danger too. Through them, foreign powers can jeopardize the security of a whole country.

Threats to democracy

The Cambridge Analytica scandal illustrates how privacy losses can contribute to the gerrymandering of democracy. Privacy violations enabled the construction of profiles that were used to target people with propaganda that matched their psychological tendencies. Christopher Wylie, the Cambridge Analytica whistle-blower, believes Brexit would not have won in the referendum if the data firm had not interfered.[33] In a way, the company harmed all of the citizens of the countries it meddled in, and citizens from other countries too, given that we are all affected by global politics. That's how far-reaching data harms can be.

Chris Sumner, research director and co-founder of the not-for-profit Online Privacy Foundation, led a research project on dark ads. So-called 'dark ads' are visible only to the publisher of the ad and the intended target of the ad. Groups can be targeted using location data, behavioural data, and psychographic information (psychographic profiling classifies people according to personality types based on personal data). Sumner set out to test how effective such targeting can be. He and his research partner, Matthew Shearing, evaluated 2,412 people's propensity for authoritarianism on Facebook, and divided them into two groups: those with high and low authoritarian tendencies. Authoritarian personalities are characterized by a tendency to obey and respect people in a position of authority, they value traditions and norms more, and they are less tolerant of outgroups. Sumner and Shearing then created advertisements that either supported or opposed state mass surveillance.

The team created four different ad campaigns. The pro-surveillance ad designed for people high in authoritarianism showed a picture of bombed buildings and read: 'Terrorists – Don't let them hide online. Say yes to mass surveillance.' The version created for people with low authoritarian tendencies read: 'Crime doesn't stop where the Internet starts. Say yes to surveillance.' On the other side, the anti-surveillance ad tailored to people with high levels of authoritarianism carried an image of the D-Day landings and read: 'They fought for your freedom. Don't give it away! Say no to mass surveillance.' The version tailored to people with low levels of authoritarianism showed a photograph of Anne Frank and read: 'Do you really have nothing to fear if you have nothing to hide? Say no to state surveillance.'

Tailored ads were more effective with their intended target groups. The ad that supported surveillance and targeted high authoritarian personalities, for instance, had twenty times as many 'likes' and shares from the high-authoritarianism group than from the low one. People classified as highly authoritarian were significantly more likely to share an ad designed for them, and people classified as having low levels of authoritarian tendencies thought that ads designed for them were more persuasive than those ads that were designed for their opposites.[34] What is not clear, however, is how these metrics (likelihood of sharing a post, and finding an ad persuasive) translate into votes.

Some sceptics have argued that microtargeting has limited effects, and that therefore we should not worry about its impact on elections. One important challenge faced by campaigns trying to influence people's opinions is that the correlation between personality traits and political values is not always a strong one. If campaigns make mistakes about people, they might mistarget messages and suffer a backlash. A further issue is that the predictive power of Facebook 'likes' has an expiry date – what you liked five years ago may not be what you like now, but you never went back to click 'unlike'. Furthermore, what 'liking' something means today might differ from what it means a year from now. 'Liking' a politician before and after an important political event such as the Brexit referendum might signal very different political stances. Political campaigns also face competition from other campaigns using the same tactics, so that at least some of the effects might cancel each other out.[35]

The concern, however, is that microtargeting does have an impact, even if limited. Research suggests, for example, that when a voter is targeted with content from an opposing party that emphasizes a particular issue in which the voter and the political candidate agree, that person is more likely to vote for the opposing party or abstain from voting altogether.[36]

When the people exposed to targeted propaganda are in the millions, you don't need to have a big effect to sway an election. In 2012, Facebook published in *Nature* the results of a randomized controlled study on 61 million users in the United States during the 2010 congressional elections. (In keeping with Facebook's style, the study seems to have been carried out without people's informed consent.)[37] On the day of the elections, one group was shown a statement at the top of their newsfeeds encouraging them to vote, along with an 'I Voted' button. Another group was shown the same statement and up to six profile photographs of the user's Facebook friends who had already clicked the 'I Voted' button. A third control group did not receive any message. The results show that those who got the message with photographs of their friends were 0.4 per cent more likely to vote. That may not seem like a big difference, but when millions of people get exposed to an influencing message or image, the numbers add up. The authors of the study claimed to have increased turnout by about 340,000 votes.[38]

If you think about how many elections are won by frighteningly few votes, 340,000 votes seem more than enough to change the course of an election. In the United States, Trump won the 2016 election by a narrow margin of 70,000 votes in three swing states.[39] In the Brexit referendum, Leave won by less than 4 per cent of the vote. Facebook encouraging all its users to vote might not be a bad thing. But what if they encourage only some people to vote and not others? And not random people, but people who are likely to vote for one particular candidate? One of the objectives of Cambridge Analytica was to identify voters they called 'persuadables' – those who could be convinced either to refrain from voting or to vote for a candidate they might otherwise not have voted for. To some people they showed fake news about the candidate they were trying to undermine, to others they showed content discouraging them from voting, and so on.

A recent documentary by Channel 4 showed how the Trump 2016 campaign categorized more than 3.5 million black Americans for deterrence. The documentary describes the kind of data the campaign had on almost 200 million voters: 'whether you own a dog or a gun, whether you're likely to get married, or planning a baby; there was even a score for personality type'. Judging by the collapse in turnout in crucial states, the documentary suggests the campaign might have been successful, although other factors came in to play. Watching black American citizens being shown how they had been targeted for deterrence, being confronted with the data that the campaign had on them, is unsettling. Many of the ads designed for voter suppression were dark ads on Facebook. According to the documentary, the Trump campaign spent $44 million to issue almost 6 million *different* ads on the platform. Facebook keeps the content of those ads a secret.[40]

That a company like Facebook has the power to influence voters should worry us. As the authors of the Facebook study point out themselves, the American presidential race in 2000 between Al Gore and George W. Bush was won by just 537 votes in Florida – less than 0.01 per cent of votes cast in that state. If Facebook had encouraged Democrat voters in Florida to go to the ballot box, and had not done the same for Republicans, Al Gore would likely have become President, and history might have taken a completely different turn.

Facebook vote buttons have been used in the Scottish referendum in 2014, the Irish referendum in 2015, the UK election later that year, the Brexit referendum in 2016, the 2016 US election, in Germany's 2017 federal elections, and in the 2017 parliamentary elections in Iceland. At least in Iceland, not all citizens were shown the buttons, but we don't know how many people saw the button and what criteria were used to decide who got to see the voting message. *We simply don't know what effect these messages have had on our elections.* Facebook keeps that information to itself.[41] Having one of the most powerful corporations in the world know so much about us and allowing it to show us messages that can influence our voting behaviour during elections is insane. Particularly if we don't even audit it. The possibility of gerrymandering democracy should be taken more seriously.

One of the most important pillars of a healthy democracy is having fair elections. Not only that, people have to feel confident about the fairness of the electoral process. If the majority of the citizenry were to suspect electoral interference, the legitimacy of the government could be seriously compromised.

There were a few reasons to worry about electoral interference and Facebook in the 2020 American race for the presidency. First, almost 70 per cent of American adults use Facebook.[42] The social network has the potential to influence the majority of American voters.* Second, Facebook has proved to be notoriously untrustworthy, as we have discussed throughout this book, failing its users so many times that keeping count has become challenging.[43]

Third, although Facebook made some desirable changes towards moderating political ads, it's still the case that their policies are constantly changing, self-imposed, self-supervised, and controversial.[44] Facebook has not only allowed lies and fake news, it has prioritized them, given that paid ads get access to tools, such as microtargeting, that maximize influence.[45]

Modern democracies have not yet developed well-established policies, mandated from the government and independently

---

* YouTube, owned by Google, and also in the business of influencing people through their data, is the only major social media platform that is used by more Americans (73 per cent).

supervised, to regulate political campaigns on all social media platforms. Facebook cannot fix itself, because it has a noxious business model according to which it's in its best interest to have inflammatory content that engages users for longer.[46]

The digital economy kindles fake news. Less than one month before the elections, research from the German Marshall Fund of the United States, a think tank, suggested that 'the level of engagement with articles from outlets that repeatedly publish verifiably false content' on Facebook had increased 102 per cent since the run-up to the 2016 election.[47]

Fourth and most important, Facebook had a private stake in who won these elections because they wanted to avoid regulation, which made it all the more tempting for them to interfere.

The company's recruitment of former Republicans in senior positions and its self-interest in staying unregulated gave rise to worries of conservative bias.[48] The fact of the matter is that, if it had wanted to, Facebook could have interfered with the elections without facing any accountability for it. Or it could have allowed other actors, such as Russian parties, to interfere with the elections. We might not even have known it had happened, unless someone blew the whistle or there was a serious investigation.[49]

Facebook may well have acted responsibly and refrained from interfering with the 2020 elections, even against its self-interest of avoiding regulation. But we shouldn't have to trust Facebook or other companies to respect our democratic processes. The rule of law cannot rely solely on good faith. Democracy can only be robust if no one is able to interfere with elections with impunity.

Watching platforms like Twitter and Facebook come up with new policies to try to curb the concerning developments of the 2020 elections was disquieting. It felt like what we thought was an experienced and robust democracy was having to relearn on the fly how to guarantee safe and fair elections. Many of the policies implemented by social media platforms seemed like haphazard patches, improvised responses to situations as they arose. More worryingly, there was something uncanny in watching private companies come up with rules to try to protect democracy – platforms that have hitherto contributed to the erosion of democracy. Twitter and Facebook do not seem like the kind of institutions that have either the expertise or the legitimacy to come up with the rules that should shape democratic elections. Robust democracies need more stable, reliable, legitimate, and transparent tools to guarantee safe and fair elections. And those rules cannot be set by companies, irrespective of how well-intentioned they might be. Of course, in the absence of legitimate and society-wide democratic rules for political campaigns online, having social media platforms try to limit misinformation and possible electoral interference is better than nothing, but it's not good enough. We cannot leave our democracies in the hands of private corporations that are as likely to undermine democracy as they are to help it, if it is in their financial interest.

We cannot allow what happened with Cambridge Analytica to happen again. Even if it is uncertain to what extent Cambridge Analytica and other similar efforts have been successful in influencing elections, what is crystal clear is that Cambridge Analytica's intent was to thwart democracy.[50] They wanted to hack the electorate. They weren't trying to disseminate true information and give good arguments for why we ought to vote for a candidate. Rather, they appealed to people's most primal emotions and were unscrupulous with the truth, showing very different content to very different people. Discouraging people from going to vote because they might support the candidate you're trying to beat is thwarting democracy. It's foul play. It is quite possible, perhaps even likely, that both Brexit and Trump in 2016 would have lost were it not for personalized political ads on social media. But even if that is not the case, attempts to hack democracy must be stopped, just like murder attempts must be stopped even if there's a chance they might be unsuccessful.

You might wonder what the difference is, if any, between microtargeted political ads based on personal data and old-fashioned political ads. After all, the digital age did not invent propaganda or false political messages. What is new and destructive is showing each person different and potentially contradictory information. Data firms try to exploit our personality traits to tell us what we want to hear, or what we need to hear to behave in the way they want us to. A candidate could get away with giving one image to some citizens and giving an opposite image to a different set of citizens.

Personalized ads fracture the public sphere into separate parallel realities. If each of us lives in a different reality because we are exposed to dramatically different content, what chance do we stand of having healthy political debates? When politicians have to design one ad for the whole of the population, they tend to be more reasonable, to appeal to arguments that a majority of people are likely to support. Personalized ads are more likely to be extreme.

When we all see the same ads, we can discuss them. Journalists, academics, and political opponents can fact-check and criticize them. Researchers can try to measure their impact. At the moment, data about political ads and campaigns is proprietary, which makes it hard or impossible for researchers to access it.[51] Scrutiny puts pressure on political candidates to be consistent. Furthermore, when ads are public, we can more easily keep tabs on political parties not spending more than the amount they are allowed, and not advertising in ways that are off-limits by law. Political advertising is highly regulated in other media such as television and radio. In the United Kingdom it is mostly banned, except for a limited number of boring 'party political broadcasts'. We can only regulate ads if we can see them, which is why dark and personalized ads have to go (more on ads in the next chapter).

When social media platforms ask us to share our data in an effort to categorize us as old or young, men or women, conservative or liberals, white or black, for or against immigration, for or against abortion, and treat us accordingly, they create and entrench divisions. We should not allow the public sphere to splinter along the lines that make us different. It doesn't have to be that way. For the public sphere to be comfortable for all, to coexist in harmony despite our differences, for pluralism to be possible, our collective life must retain a degree of neutrality, for which we need liberalism.

Threats to liberalism

The basic tenet of liberal societies is that individuals should have the freedom to live their lives as they see fit. There is a presumption in favour of liberty such that 'the burden of proof is supposed to be with those who [. . .] contend for any restriction or prohibition', as John Stuart Mill wrote.[52] Rules should be put in place to avoid harm to people, to ensure citizens are free from unnecessary interferences, and to establish a common life in which all can participate.

Privacy is important to build a robust private sphere, a bubble of protection from society in which individuals can enjoy times and places free from others' gazes, judgements, questions, and intrusions. Privacy norms serve a valuable function in giving us breathing space. A healthy degree of reticence and concealment is necessary for civilized life to run smoothly.[53] If we could all read each other's minds at all times, the private sphere would shrink into nothingness, and the public sphere would become contaminated with endless unnecessary conflicts. Liberalism is not only about governments not interfering with citizens' private lives. For liberalism to thrive, it must be embedded in a similar culture of restraint in which ordinary citizens make an effort to let one another be.

Social media encourages us to 'share' online. Facebook's business model depends on people revealing aspects of themselves online. When users share less personal content, Facebook worries and tweaks the platform to encourage more sharing.[54] Share everything you can, is the message. Tell us who you are, tell us how you feel, tell us about your family and friends, tell the world what you think about other people. We want to know. We want to hear what you have to say.

Social media platforms promote a culture in which people are discouraged from holding back. The more people share, the more data can be analysed and used to sell access to us. The more comments people make on what others share, the more clicks, the more ads, the more money, the more power. It might seem like a win-win situation. We get to rant endlessly, and tech companies get to earn their keep. Except that much of what gets shared online does not benefit users. As a user, your data gets exploited in ways that are not in your interest, and what you share exposes you to other users, some of whom are only too happy to troll, blackmail, or shame you. Social media is communication unrestrained. But civility requires a measure of restraint – restraint in what you share about yourself, in the opinions you express (especially about others), and in the questions you ask.

Restraint need not amount to dishonesty. Just as clothing does not mislead others about you being naked underneath, not expressing how stupid you think another person is does not amount to deceit. We do not need to know everything about each other to have a frank conversation. You don't need to know another person's darkest fears, secrets, and fantasies to be friends with them, much less good neighbours. We don't want to tell our fellow citizens everything about us. And, just as importantly, we don't want to know everything about them.

Expecting saintliness from people in body, speech, and mind at all times is both unrealistic and unreasonable. As the philosopher

Thomas Nagel points out, 'Everyone is entitled to commit murder
in the imagination once in a while.'[55] If we push people to share
more than they otherwise would, we will end up with a more pernicious social environment than if we encourage people to curate
what they bring into the public sphere. A culture of exposure
impels us to share our imaginary acts of murder, and in so doing
needlessly pits us against each other. Sparing each other from
our less agreeable sides is not a fault – it's a kindness.

Liberalism asks that nothing more should be subjected to
public scrutiny than what is necessary to protect individuals and
cultivate a wholesome collective life. A culture of exposure
requires that *everything* be shared and subjected to public inspec-
tion. Big tech sells the fantasy that those who do nothing wrong
have nothing to hide, that transparency is always a virtue. It's not.
Exhibitionists who flash other people are not being virtuous. In
the digital economy, everyone is pushed to express more than is
necessary for the purposes of friendship, effective communication, and public debate – all in an effort to create more data.

Oversharing benefits big tech, not users. It makes the public
sphere uninhabitable. Such relentless social pressure to share leads
to expressions of aggression and intolerance, and to vigilantism
and witch-hunting. There is no truce. Every image, word, and
click gets collected and monetized by companies, and scrutinized
and potentially torn apart in a cathartic public act of online shaming by netizens. Such constant fuss about every detail of what
people say and do distracts us from more important conversations –
from talking about issues of justice, economics, ecology, and
public goods. While we are busy endlessly bickering online,

trolling each other, and tearing people to shreds for human weaknesses that we probably share, our democracies are falling apart.

In some ways, cultures of exposure resemble the brutality of children's social relations. Children are as notorious for not knowing when to stop talking as they are for their potential cruelty – especially when they are in a group. Perhaps as the internet matures we will distance ourselves from a culture of oversharing and bullying, and approach more adult ways of relating to one another.

Threats to the safety of individuals

Personal data can be, is, and will continue to be misused. And some of the abuses of personal data are deadlier than asbestos.

One of the most lethal misuses of data was carried out by the Nazi regime during the Second World War. When Nazis invaded a country, they were quick to take hold of the local registries as a first step to controlling the population, and in particular, finding Jews. Countries varied widely in both the kind of records they held and their reaction to Nazis' thirst for data. The starkest comparison is that between the Netherlands and France.[56]

Jacobus Lambertus Lentz was not a Nazi, but he did more for the Nazi regime than most zealous anti-Semites. He was the Dutch Inspector of Population Registries, and he had a weakness for population statistics. His motto was: 'to record is to serve'. In March 1940, two months before the Nazi invasion, he proposed a personal identity system to the Dutch government in which all citizens would be required to carry an ID card. The card used translucent inks that disappeared under a quartz lamp and watermarked paper to make sure it could not be forged. The government rejected his proposal, arguing that such a system was contrary to Dutch democratic traditions, as it would treat ordinary people like criminals. Lentz was deeply disappointed. A few months later he would propose the same idea to the Reich Criminal Police Office. The occupied forces eagerly put it into practice. Every Dutch adult was required to carry an identification card. A *J* was stamped on those cards carried by Jews – a death sentence in their pockets.

In addition to these cards, Lentz used Hollerith machines – tabulating machines sold by IBM that used punch cards to record and process data – to expand records kept on the population. In 1941, a decree was issued requiring all Jews to register at their local Census Office. For decades the Dutch had naively collected data on religion and other personal details, aspiring to have a comprehensive system that could follow each person 'from cradle to grave'. Lentz and his team used the Hollerith machines and all the information available to them to make it easier for Nazis to track down people.

In contrast to the Netherlands, censuses in France did not collect information about religion for privacy reasons. The last census to collect such data was taken in 1872. Henri Bunle, the chief of the General Statistics Office of France, made it clear to the General Commission for Jewish Questions in 1941 that France did not know how many Jews it had, let alone where they lived. Furthermore, France did not have an extensive punch card infrastructure like that of the Netherlands, making it hard to collect new data. If

Nazis wanted police departments to register people, they would have to do it manually, on pieces of paper and index cards.

Without Hollerith machines there was no way of sorting and tallying the information being collected about the population. The Nazis were desperate. René Carmille, Comptroller General of the French Army, and an enthusiast for punch cards who owned tabulating machines, including Hollerith ones, volunteered to bring order to the chaos and deliver the Jews of France to their executioners.

Carmille developed a national Personal Identification Number that worked as a descriptive bar code for people; it is the precursor of France's current social security number. Different numbers were assigned to represent personal characteristics such as profession. Carmille also prepared the 1941 census for all French citizens aged fourteen to sixty-five. Question 11 asked Jews to identify themselves through both their grandparents and their professed religion.

Months passed, and the lists of Jews expected from Carmille did not arrive. The Nazis grew impatient. They started rounding up Jews in Paris, but without Carmille's tabulations they mostly relied on Jews turning themselves in. More months passed, and still the lists did not come.

Unbeknown to the Nazis, René Carmille had never intended to hand over his fellow citizens. He was one of the highest-placed operatives of the French resistance. His operation generated some 20,000 fake identities, and he used his tabulation machines to identify people who could fight against the Nazis. The answers to question 11 regarding whether people were Jewish were never tabulated. Those holes were never punched, that data forever lost. Over 100,000 altered punch cards have been found – punch cards never surrendered to the Nazis. Hundreds of thousands of people saved by just *one* person who decided *not* to collect their data – their toxic data.

It seems reasonable to suppose that Carmille knew he would eventually get caught if he didn't surrender the data he promised. He was discovered and detained by the SS in 1944. He was tortured for two days, and was then sent to Dachau, where he died from exhaustion in 1945.

Data collection can kill. The Dutch had the highest death rate of Jews in occupied Europe – 73 per cent. Of an estimated 140,000 Dutch Jews, more than 107,000 were deported, and 102,000 of those were murdered. The death rate of Jews in France was 25 per cent. Of an estimated 300,000 to 350,000 Jews, 85,000 were deported, and 82,000 of those were killed. A lack of privacy was instrumental in the killing of hundreds of thousands of people in the Netherlands, while privacy saved hundreds of thousands of lives in France. Further support for the hypothesis that it was data collection that made the difference between the two countries relies on Jewish refugees living in the Netherlands experiencing an overall death rate lower than that of Dutch Jews. Refugees had avoided registration.[57]

Other documented instances of misuse of personal data include the removal of American Indians from their lands in the United States in the nineteenth century; the forced migration of minority populations in the Soviet Union in the 1920s and 1930s; and the use of a registration system implemented by

Belgians in the 1930s to find and murder Tutsis during the Rwanda genocide of 1994.[58]

The best predictor that something will happen in the future is that it has happened in the past. These stories are not from a distant galaxy in a world of fiction. These are true stories that we must learn from in order to avoid repeating the deadly mistakes of the past.*

Imagine a contemporary authoritarian regime getting hold of all our personal data. Despots from the past have had scraps of data in comparison to the thousands of data points one can acquire about anyone in the world today with just a few clicks. An authoritarian government could learn about our every weakness inside out without much effort. If it could predict our every move, it could be the beginning of an undefeatable regime. To get a sense of just how dangerous personal data is, imagine a contemporary regime similar to the Nazis, having real-time data of your location, your face print, your gait, your heartbeat, your political beliefs, religious background, and much more.

Among the many data-related accounts from the Second World War, one stands out as particularly instructive. In March 1943, a cell of the Dutch resistance attacked the municipal registry of Amsterdam. The objective was to destroy as many records as possible in an attempt to help 70,000 Jews in Amsterdam escape murder. Gerrit van der Veen, Willem Arondéus, Johan

---

* In view of this dark history of ID cards, the United Kingdom's decision to abolish them in 1952 is understandable, as is the reluctance in recent debates to revive them.

Brouwer, Rudi Bloemgarten and others entered the building dressed as policemen. They sedated the guards, sparing their lives, soaked the files in benzene, and set the documents on fire. Sympathizers among the firefighters knew of the attack. When the alarm sounded, they made an effort to delay the deployment of the trucks, giving the flames time to do their job. When they reached the registry, they used as much water as possible to damage as many records as possible.

Unfortunately, the attack on the registry was not very successful. Twelve members of the resistance cell were found and executed. And the fire destroyed only about 15 per cent of the documents.[59]

Just as the Nazis knew to go to the registries, today's wrongdoers know where to find our data. And they don't even have to invade a country with troops to get hold of our most sensitive information. They just need a good hacker. In that sense, the risk to our personal data, and everything that our privacy protects, is much higher than in the pre-internet world.

We should learn from the mistakes of the past. Personal data is toxic, and we should regulate it as such. Let's not do what we did with asbestos. We put asbestos everywhere: in car brake linings, pipes, ceiling and floor tiles, concrete, cement, bricks, clothing, mattresses, electric blankets, heaters, toasters, ironing boards, cigarette filters, and artificial snow, among other things. Once it was in our roofs and walls, in the very structure of the places we inhabit, it became very difficult to extract it without risk. Asbestos kills hundreds of thousands of people every year.

It is still poisoning people around the world, even in places where it has now been banned.[60]

Let's not allow personal data to poison individuals, institutions, and societies. Luckily for us, it is not too late to correct our current trajectory when it comes to personal data. We can fix the internet and the economy. Let's learn from the experience of the Netherlands during the Second World War. The Dutch made at least two big mistakes with respect to privacy. They amassed too much personal data. And once they realized how toxic that data was, they didn't have an easy and fast way to delete it. We are making both of those mistakes at an unprecedented scale. We need to change that before it's too late.

FIVE

PULLING THE PLUG

The surveillance economy has gone too far. It has abused our personal data in too many ways, too many times. And the quantity and sensitivity of the data being traded makes this grand experiment too dangerous to be continued. We have to put a stop to the trade in personal data.

The data economy has to go because it is at odds with free, equal, stable, and liberal democracies. We can wait for a truly massive data disaster before we start to protect privacy – anything from a monumental leak of biometric data (consider that, unlike passwords, our faces are not something we can change) to the misuse of personal data for the purposes of genocide – or we can reform the data economy now, before it's too late.

Personal data has become such a big part of the economy that it might sound unrealistic to pull the plug on it. But once upon a time the idea of recognizing workers' rights sounded just as outlandish, if not more so. Today, we look at the past and lament the savageness of exploitative labour practices, for instance, during the industrial revolution. Tomorrow, we will look at today and lament the foolishness of the surveillance economy.

Although human beings do not always excel at averting disaster, some examples show that we are capable of coordinating our actions and redirecting an ill-fated path. Ozone in the outermost layers of the atmosphere absorbs most of the sun's ultraviolet rays. Without an ozone layer to protect us, our eyes, skin, immune system, and genes would get damaged by ultraviolet rays. As the ozone layer thinned in the second half of the twentieth century, the incidence of skin cancers went up. In 1985, a group of scientists published an article in *Nature* describing the extent of the annual depletion of the ozone layer above the Antarctic. We were headed for disaster.

Only two years later, in 1987, the Montreal Protocol was signed, an international agreement aimed at banning the production and use of ozone-damaging chemicals, including CFCs (chlorofluorocarbons). These chemical compounds were used worldwide in refrigerators, air conditioners, and aerosol cans. What made them attractive was their low toxicity, flammability (like asbestos), and reactivity. Unfortunately, the fact that they don't react with other compounds also makes them dangerous, because it gives them a long lifetime during which they can diffuse into the atmosphere.

Thanks to the opposition of experts and the public to the

production and use of CFCs, industry innovated and found alternatives. Ozone holes and thinning have been recovering at a rate of about 1 to 3 per cent a decade since 2000. At this rate, the ozone layer over the northern hemisphere is expected to be completely healed by the 2030s. By 2060, ozone will have made a full comeback worldwide. Phasing out CFCs had a further benefit: it halved global warming.[1]

If we can save the ozone layer, we can save our privacy.

Most of the recommendations in this chapter are aimed at policymakers. Much like saving the ozone layer, ending the data economy requires regulation. There is no way around that. But what will make policymakers act is pressure from *you*, from us, from the people. It is ultimately up to us to demand an end to the personal data trade, and there is much you can do to help that effort.

Policymakers are often eager to protect us. But they may fear the consequences of taking a bold step – maybe their party colleagues will disagree, maybe voters will not appreciate what is being done on their behalf, maybe it'll hurt their chances of climbing the political ladder. Politicians derive their power from us. If they know we care about privacy, and that we will withdraw our votes and support if they do not regulate for privacy, you can be sure they will act. They are just waiting for our cue. Our job is to be as well informed as possible so that we know what to ask of our politicians. You can express your convictions by getting in touch with your representatives, voting, and by protecting your privacy, which is the topic of the next and final chapter.

STOP PERSONALIZED ADVERTISING

Let's go back to where we started. The origin of the dark sides of the data economy is in the development of personalized advertising – therein lies the beginning of a solution. Micro-targeted ads that are based on your identity and behaviour are not worth the negative consequences they create.

One of the gravest dangers of personalized advertising, as we saw when we discussed the toxicity of personal data, is the possibility that it might corrode political processes. You might think that a more reasonable solution to that problem is to ban political ads, as Twitter did in 2019. But it is not easy to demarcate clearly what is political from what is not. Twitter defines political content as 'content that references a candidate, political party, elected or appointed government official, election, referendum, ballot measure, legislation, regulation, directive, or judicial outcome'. What about ads denying climate change? Or informing the public about climate change? Or ads against immigration? Or ads that advertise family planning health centres? All these seem political, and could be intimately related to a particular candidate or election, and yet it is not clear that they would or should be banned by Twitter.

A better solution is to ban personalized ads completely. It's not just that they polarize politics, they are also much more invasive than most people realize. When you see a personalized ad, it doesn't just mean that a given company knows more about you than your friends do. It's much worse than that. As the page loads, and many times before you even get a chance to consent (or not) to data collection, competing advertisers bid against each other for the privilege of showing you their ad.

Real-time bidding (RTB) sends your personal data to interested advertisers, often without your permission. Suppose Amazon gets that data and recognizes you as a user who has visited their website before in search of shoes. They might be willing to pay more than others to lure you into buying shoes. And that's how you get shown an Amazon shoe ad. Unfortunately, in that process, very personal data such as sexual orientation and political affiliation might have been sent to who knows how many possible advertisers without your knowledge or consent. And those companies are holding on to your personal data.[2]

The allure of behavioural advertising is understandable. Users don't want to be exposed to products they have no interest in. If you couldn't care less about tractors, seeing tractors flash on your screen is a nuisance. In turn, advertisers don't want to waste their resources showing ads to people who would never buy their product. As the nineteenth-century retailer John Wanamaker famously said, 'Half the money I spend on advertising is wasted; the trouble is I don't know which half.'

Targeted advertising promises to solve both problems by showing customers what they are interested in buying, and making sure advertisers only pay for ads that will increase their sales. That's the theory, a win-win situation. Unfortunately, the practice looks nothing like the theory. The practice has normalized surveillance. It has led to the spread of fake news and clickbait. It has fractured the public sphere, and it has even compromised our democratic processes. As if all of these externalities weren't enough, microtargeted advertising doesn't even do what it says on the tin: it doesn't show us what we want to see, and it's not clear that it either allows advertisers to save money or increases their sales.

Advertising is, for the most part, a less scientific endeavour than one might imagine. Marketers often pursue an advertising strategy more out of a gut feeling than because they have hard evidence about what is going to work. In some cases, this intuitive approach has led to prominent businesses wasting millions of pounds.[3]

There is not enough research to allow us to assess the effectiveness of targeted advertising with a high degree of confidence. There is, nonetheless, reason to think personalized ads are not as profitable as optimists had hoped.[4] Preliminary research shows that advertising using cookies does increase revenues, but only by around 4 per cent – an average increase of just $0.00008 per ad. Yet advertisers are willing to pay much more for a targeted ad than for a non-targeted one. According to one account, an online ad that doesn't use cookies sells for just 2 per cent of the cost of the same ad with a cookie.[5] 'There is a sort of magical thinking happening when it comes to targeted advertising [that claims] everyone benefits from this,' says Alessandro Acquisti, Professor at Carnegie Mellon University and one of the authors of the study. 'Now at first glance this seems plausible. The problem is that upon further inspection you find there is very little empirical validation of these claims.'[6]

If targeted ads are much more expensive than non-targeted

ads, and the increase in revenues they offer is marginal, we may
be losing our privacy for nothing at all. Platforms such as Google
and Facebook might be unduly profiting from selling smoke.* A
poll by Digiday confirms this suspicion. Out of forty publishing
executives who participated in the survey, for 45 per cent of
them, behavioural ad targeting had not produced any notable
benefit, while 23 per cent of respondents said it had actually
caused their ad revenues to decline.[7] In response to the GDPR,
the *New York Times* blocked personalized ads yet did not see ad
revenues drop; rather, they rose.[8]

One reason targeted ads may not be very successful in
increasing revenue is because people hate them.[9] Do you
remember when ads were creative and witty? Ads used to be
interesting enough that you could compile them in a one-hour
TV show and people would *want* to watch them. Not any more.
Most ads these days – especially online ads – are unpleasant at
best and abhorrent at worst. They are typically ugly, distracting, and intrusive. Contemporary advertising has forgotten the
lessons of David Ogilvy, known as the father of advertising,
who wrote that 'you cannot *bore* people into buying your product; you can only *interest* them in buying it'. You cannot (and
should not) *bully* people into buying your product either: 'It is
easier to sell [to] people with a friendly handshake than by

---

* It is worth bearing in mind, though, that even if targeted ads are
  not worth their cost, big platforms can provide access to such a
  large audience, that it might still be in the advertisers' interests
  overall to use those platforms.

hitting them over the head with a hammer. You should try to *charm* the consumer,' Ogilvy wrote.[10] In some ways, online ads are worse than being hit by a hammer.

People may especially hate targeted ads because they invade our privacy. Have you ever felt inappropriately watched by your ads? You tell a friend about a sensitive topic – perhaps you're thinking of changing jobs, or having a baby, or buying a house – and the next ad you see is directly related to what you thought was a private conversation. Unsurprisingly, research suggests that ads are less effective when people consider them creepy.[11] If people know that an ad targeted them by tracking them across the web, or by making inferences about them, they are less likely to engage with it.

Google sensed that people would not appreciate being spied on long ago and adopted a secretive approach, as explained earlier. Do you remember the first time you started to understand how your data was being used by big tech? I'm guessing you didn't learn about it from a clear message from one of the big platforms. Maybe you started to notice how the ads you saw were related to you but different from those seen by your friends and family. Or perhaps you read about it in an article or a book.

That targeted advertising may not deliver the advantages it was designed to offer makes our loss of privacy seem all the more futile and absurd. But even if targeted ads worked in showing us what we want to see and increasing the revenue of merchants, we would still have good reason to scrap them.

Targeted ads may not work very well for businesses, but they might work quite well for swaying elections, as we've seen. A

4 per cent effect in selling a product will not be enough to compensate for the cost of the ad, but that same effect in terms of numbers of voters could very well decide an election.

Personalized ads have normalized hostile uses of tech. They have weaponized marketing by spreading misinformation, and they have shattered and polarized the public sphere. As long as platforms like Facebook use personalized ads, they will remain divisive by exposing us to content that pits us against one another, despite the company's mission statement to 'bring the world closer together'. Facebook will be all the more damaging as long as it dominates online advertising.

Facebook takes publishers away from their own distribution channels and encourages clickbait content. That the relationship between publishers and their audiences is weakened is especially troublesome in the case of newspapers. It makes them dependent on platforms that can change their algorithm and hurt their visibility.[12] Even before Facebook announced a change in its algorithm in 2018 to promote posts by family and friends, as opposed to content produced by publishers, news organizations were already experiencing a dive in Facebook-referred traffic. Anecdotally, some sites reported a 40 per cent drop. BuzzFeed had to fire people, and the biggest newspaper in Brazil, *Folha de S.Paulo*, pulled its content from Facebook.[13]

Banning targeted advertising will boost competition. One of the elements that is preventing competition against Facebook and Google is the amount of personal data they have hoarded. Everyone wants to advertise with them partly because there is an assumption that the more data a platform has, the more effective it will be at personalizing ads. If everyone used contextual advertising instead, platforms would be on a more equal footing.[14] Contextual advertising shows you ads of shoes when you type 'shoes' in your search. It doesn't need to know who you are or where you've been. If companies were not allowed to use personal data for ads, that would do away with some of the competitive advantage of Google and Facebook, although the two tech giants would still be advertising behemoths given the number of users they have.

There is a place for advertising in the online world. Especially for informative advertising (as opposed to combative or persuasive advertising), which, in David Ogilvy's view, is the kind of marketing that is both more moral and more profitable. Online advertisers would do well to remember Ogilvy's adage that 'advertising is a business of *words*'. Perhaps online ads should look more like magazine ads than television ads. Instead of designing noxious ads that both surveil and distract us out of our minds with jumping flashy images, online ads could strive to be based on words and facts, following Ogilvy's ideal. Including facts about a product, as opposed to adjectives, and adding good advice, such as how to remove a stain, or a food recipe, are examples of good practices.[15] Online advertisers should offer us information, instead of taking information from us.

Ads are particularly justified in the case of new products and brands. But they do not have to violate our right to privacy to be effective. Furthermore, there is an argument to be made for limiting the share of the economy that is dedicated to advertisement. At the moment, ads are the core of the data economy. Having too many ads dominating our landscape, however, might be bad for wellbeing.

A recent study of approximately one million European citizens across twenty-seven countries and over three decades suggests that there is a correlation between increases in national advertising expenditure and declines in levels of life satisfaction. After taking account of other macroeconomic variables such as unemployment and individuals' socioeconomic characteristics, researchers estimate that a national doubling of ad spending is associated with a subsequent drop in reported satisfaction of 3 per cent – an effect about a quarter as strong as that of being unemployed.[16] If advertising is boosting our economy at the cost of our happiness, we might want to think twice about the kind of footing we allow it to have in our lives.

According to a report commissioned by the Association of National Advertisers and The Advertising Coalition, advertising comprised 19 per cent of the United States' total economic output in 2014.[17] To put that into perspective, tourism contributed 7.7 per cent the same year.[18] The American advertising market value is larger than the banking industry.[19] And yet this is an industry that, it seems, makes us miserable. Like former Facebook data scientist Jeff Hammerbacher, I also find it depressing that 'the best minds of [our] generation are thinking about how to make people click ads'.[20]

Limiting ads would also be a natural way of curbing the power of big tech platforms that heavily depend on them. Let's not forget that ads constitute the vast majority of Alphabet's and Facebook's revenue.[21]

Personalized ads have to stop. Real-time bidding should be banned. We should limit the dominance of ads, or modify them so that they don't have a negative effect on people's wellbeing. Luckily, you don't have to wait around for policymakers to reform the advertising industry: you can use adblockers (see next chapter for details).

STOP THE TRADE IN PERSONAL DATA

Personal data should not be something you can buy, sell, or share to exploit for profit. The opportunities for abuse are too many, and proliferating. The more sensitive the data, the stricter the ban and the steeper the penalty for breaking the law ought to be. That we are allowing companies to profit from the knowledge that someone has a disease, or has lost their son in a car accident, or has been the victim of a rape, is revolting.

I have never encountered a good argument to justify data brokers. Data brokers are the scavengers of the digital landscape. They live off the data trails we leave behind, sell them to the highest bidder, and very rarely have any regard for the people whose data they are profiting from.

Twenty years ago, Amy Boyer was murdered by her stalker after he purchased her personal information and location data from Docusearch[22] – a data broker that, incredibly, still exists. On their website they claim to be 'online & trusted for over 20 years'. Data vultures can't be trusted. Data brokers have sold people's data to fraudsters. In 2014, LeapLab, a data broker in

Nevada, sold intimate details of hundreds of thousands of people to a 'company' that used those records to make unauthorized withdrawals from people's bank accounts.[23] Have you ever had money disappear from your account? You might have a data broker to thank for that; they might've sold or lost your data. Equifax's data breach, discussed in the previous chapter, is one of the worst in corporate history.[24] That data-related tragedies have been relatively few, given the extensive neglect of data security, is a testament to humans being generally law-abiding and decent. But we can't always rely on the kindness of people. We need better security measures.

The very existence of sensitive files on internet users is a population-level risk. Many times, personal data held by data brokers is not even encrypted or well protected. Data brokers currently don't have enough of an incentive to invest in good security. Foreign governments and malicious actors can hack that data and use that knowledge against us. The more data brokers collect our personal details, and the more they sell those files to yet further companies, our risk of suffering harm from a data breach grows. And what do we get in return? Nothing. Were we drunk when we struck this deal? No, we were never asked.

Buying profiles from data brokers is not even expensive. Bank account numbers can be bought for 50 cents, and a full report on a person can cost as little as 95 cents.[25] For less than $25 per month you could run background checks on everyone you know (but please don't). In May 2017, Tactical Tech and artist Joana Moll purchased a million online dating profiles from USDate, a dating data broker. The haul included almost

5 million photographs, usernames, email addresses, details on nationality, gender and sexual orientation, personality traits, and more. Although there is some doubt regarding the source of the data, there is evidence that suggests it came from some of the most popular and widely used dating platforms. It cost them €136 (about $150).[26] That such a transaction is possible is astounding. And barbaric. Personal data being both so *valuable* and so *cheap* is the worst possible combination for privacy.

Part of what good regulation entails is stopping one kind of power turning into another. For instance, good regulation prevents economic power turning into political power (i.e. money buying votes, or politicians). In the same way, we need to stop the power accrued through personal data transforming into economic or political power. Personal data should benefit citizens – it shouldn't line the pockets of data vultures.

Even in the most capitalist of societies we agree that certain things are not for sale – among them are people, votes, organs, and the outcomes of sports matches. We should add personal data to that list. Personal data sounds too abstract. This abstraction is all too convenient for data vultures. What we are actually talking about are our hopes and fears, our medical histories, our most private conversations, our friendships, our darkest regrets, our traumas, our joys, the way we sound and the way our heart beats when we make love*[27] – that's what's getting exploited for profit, too many times, against our best interests.

---

* If you have a wearable device, it is tracking, recording, and analysing your heartbeat throughout your day; sexual activity can be inferred. In

Banning the trade in personal data does not mean banning the collection or proper use of such data. Some kinds of personal data are necessary. Sharing your personal data with your doctor, for instance, is necessary to receive adequate care. But our health system should not be allowed to share that data, much less sell it.

Ending the trade in personal data does not mean that other kinds of data should not be shared – the ban need only apply to *personal* data. In fact, some non-personal data should be shared widely to promote collaboration and innovation. As computer scientist Nigel Shadbolt and economist Roger Hampson argue, the right combination is to have 'open public data' and 'secure private data'.[28]

We need, however, stricter definitions of what counts as personal data. At the moment, legislation such as the GDPR does not apply to anonymized data. As we saw in Chapter One, however, all too often data that was thought to be anonymous has ended up being easily re-identified. Part of the problem is that we are not sure what techniques may be developed and used in the future to re-identify individuals in an 'anonymous'

2019, *Bloomberg*, the *Guardian*, and *Vice News* revealed that Amazon, Google, Facebook, Microsoft, and Apple had been using human contractors to analyse voice-assistant recordings. Contractors admitted to sometimes listening to people having sex. An Apple whistleblower said: 'I heard people talking about their cancer, referring to dead relatives, religion, sexuality, pornography, politics, school, relationships, or drugs with no intention to activate Siri whatsoever.'

database. Therefore we have to be as tough as our imaginations allow us in defining what counts as anonymous.

We also need to have a very broad understanding of what counts as a data trade. Data brokers provide personal data in exchange for money, but many other companies make data deals that are less crude. Facebook, for instance, has given other companies access to its users' personal data in exchange for these companies treating Facebook favourably on their platforms. Facebook gave Netflix and Spotify the ability to read its users' private messages, and it gave Amazon access to users' names and contact information through their friends. Part of what it received in return was data to feed its invasive friend-suggestion tool, 'People You May Know'.[29] Personal data should not be part of our commercial market. It should not be sold, disclosed, transferred, or shared in any way for the purposes of profit or commercial advantage.

Again, you don't have to wait for policymakers to ban the trade in personal data to start working towards that goal if you follow the advice in the next chapter.

STOP DEFAULT PERSONAL DATA COLLECTION

Some big tech became big by plundering our data without asking for permission, without thinking about the possible consequences their actions might have for their users and society at large. This reckless attitude is best described by Facebook's

internal motto: 'Move fast and break things'. Big tech's strategy
has been to do what they please until they face resistance. Once
resistance is encountered, big tech usually tries to ignore it. When
that doesn't work, they try to seduce people with extra perks, and
to exhaust their critics with endless empty responses. Only when
resistance is persistent does big tech take a step back, and usually
after having taken many steps forward. What they hope with this
cycle, argues Shoshana Zuboff, is that we will gradually become
accustomed to accepting conditions we would never have agreed
to had they been presented to us upfront from the start.[30]

It was through this cycle that we became used to our data
being collected automatically by everyone with the means to
collect it. We have put up with it because we learned about it
years too late, once we were hooked on digital tech, and because
we were told that it was necessary for our gadgets to keep on
working as they do, and that everyone was doing it anyway. We
were also told that surveillance was needed to keep us safe.
Only when corporations faced a techlash, and regulation like
the GDPR was introduced, did they make some concessions,
such as telling us a bit about the kind of data they hold on us.
But that's not enough. Now we know better. We know that it is
possible to have cutting-edge tech gadgets without privacy
invasions. And we know that privacy is an important component of ensuring our security.

The existing state of affairs is one of ubiquitous data collection. Almost every website, app, and gadget you interact with is
collecting your data. Some of those companies don't even know
what to do with it. They collect it just in case it might be useful

in the future. But, as we have seen, data collection is not harmless. It puts us all at risk.

So far, legislation has mostly addressed the *uses* of data, not its collection. Even if the GDPR includes a principle of data minimization, according to which companies should collect only adequate, relevant, and necessary data, many institutions seem to be shielding behind a very broad interpretation of what are 'legitimate interests' for processing data. We need to be tougher on data collection.

Anyone who has ever used the internet knows that the current system of 'consenting' to data collection is faulty. It puts too heavy a burden on citizens. Not only is it annoying to have to click on sometimes dozens of options to reject data collection; because you said 'no' to cookies, you are punished by having to go through the same process every single time you use that website. If there was no data collection by default, people wouldn't need to constantly ask for their privacy to be respected. And people who opted in to data collection could legitimately be remembered, so they would only need to do it once.

The default – for companies, government agencies, and user settings on every website and app – should be *not* to collect data, or only to collect the minimum *necessary* data. Defaults matter, because most people will never touch their settings. People should have to opt in to data collection, rather than opt out.

What counts as necessary data should be understood narrowly, as data that is indispensable in order to provide a worthwhile service – not to fund that service, by selling our data or access to us through our data, but to build or maintain the service. Some services need people's data to map traffic, for instance – but they don't need *everyone's* data to do it effectively. If data from a sample of users is enough, any data collection beyond that is unnecessary.

We need to invest more in privacy innovation. If big tech is forced to face the challenge of inventing ways of using data while protecting privacy, they have a good chance of rising to the occasion. If they are allowed to continue as they have to date, such innovations might never be developed.

A promising method of collecting data is through *differential privacy*. Differential privacy essentially means inserting just enough mathematical noise into a database so that you are successful at camouflaging every member of the database – you can't infer anything in particular about any individual – but not so much as to prevent it from providing accurate responses when performing statistical analyses. It may sound complicated, but here's a simple example to illustrate the idea.

Suppose you want to know how many people in London voted for Brexit. Ordinarily you would call a few thousand telephone numbers and ask every person which way they voted. Even if you did not collect names, if you collected phone numbers and how people voted, those voters could be easily identified, and their right to a secret vote would be compromised. To collect data using differential privacy, in contrast, you would also call up a few thousand telephone numbers, but instead of asking directly how people voted, you would ask them to flip a coin. If the coin lands on heads, then people should tell you how they voted. If it lands on tails, they should flip the coin again, and if it lands on heads this time, they should tell you the truth; if it lands on tails, they should lie. Importantly, people should never tell you how the coin landed. Because you controlled how often people lied to you, you know that approximately one quarter of your results are incorrect (a lie), and you can statistically adjust for that. The result is a database that is almost as accurate as an ordinary one, and which contains no personal data, because only respondents know how the coin landed. You have no way of knowing who voted for Brexit, but you can know roughly how many people voted for Brexit. Every participant enjoys 'plausible deniability': they can claim they didn't vote for Brexit, and no one would be able to prove otherwise (at least on the basis of this database).[31]

Of course, not every kind of data can be collected using differential privacy, and the method needs refinement for institutions to be able to implement it easily and effectively. I don't mean to suggest that differential privacy is perfect or the solution to all of our problems. It's not. And if it's not implemented properly, it can create a false sense of security. Nevertheless, I still like it as an example because it illustrates that there are creative ways of analysing data without endangering people's privacy. Homomorphic encryption and federated learning are two other techniques worth exploring. We should be investing more in developing privacy tools, as opposed to investing only in methods of exploiting privacy for profit, convenience, or efficiency.

Whenever there is no alternative but to collect personal data, such data should only be collected when an individual meaningfully and freely consents to such collection, and when uses of that data and plans to delete the data have been specified (more on that last point below). Limiting the collection of personal data is not enough, however, because sensitive information can be acquired not only through data collection, but also through inferences.

STOP SURREPTITIOUS
SENSITIVE INFERENCES

Institutions thirsty for knowing more about us can escape the limits we have set for them through inferring, rather than collecting, sensitive information about us. The digital trails we leave behind as we interact with technology are routinely treated as samples of behaviour that are then used to make inferences about us.

Theories about what our data trails say about us have mushroomed in the last few years. The way people use smartphones can be used to predict test scores in cognitive abilities such as memory and concentration. Memory problems can be identified by how quickly people type on their phone, the errors they make, and how fast they scroll down their list of contacts.[32] Facebook 'likes' have been used to infer sexual orientation, ethnicity, religious and political views, personality traits, intelligence, happiness, use of addictive substances, parental separation, age, and gender.[33] Patterns of eye movements can be used to detect dyslexia. Your walking pace, measured by the accelerometer in your smartphone, can be used to infer your life expectancy. Your posts on Twitter and your facial expressions can be used to uncover depression. The list goes on, but you get the picture:

external signals are being systematically used by companies and institutions to infer private information about you.[34]

There are plenty of concerns regarding sensitive inferences, some of which are shared with other kinds of privacy-invasive practices, and some of which are particular to inferences. As is the case with surreptitious personal data collection, it is worrisome that your privacy can be violated without you ever knowing about it. Worse, you have little or no control over some of these external signals, such that there is not much you can do to protect yourself. You can try to avoid giving away your personal data, but you cannot change your face or your gait, for instance, or the way you type on your phone. Those are all involuntary markers. And there is no way for you to know whether someone is using that information and for what purposes.

A further concern about sensitive inferences is that they might be wrong about you, but they can still be used against you. Inferences based on algorithms are probabilistic – they are right only some of the time. How accurate inferences are varies widely, and companies may not have much of an incentive to make sure they are as accurate as possible. As long as companies feel that inferences give them an advantage, they may be content to use them, even if they are imperfect.

Researchers, for instance, were able to infer correctly whether a person smoked based on their Facebook 'likes' in 73 per cent of cases.[35] Suppose a company uses such an inference as a filter for hiring employees. If they have enough applicants for a job, they may not be bothered by being wrong about 27 per cent of those applicants because, from their perspective, they are still better off than if they hadn't tried to infer that information. If you are one of the unlucky people who is misclassified as a smoker, however, you have suffered an injustice, and you may never know about it, because the company might never tell you why you didn't get the job.

In some cases, sensitive inferences may be acceptable. As a patient, you may want your doctor to analyse how you type on your smartphone so that they can detect possible cognitive problems as early as possible. But sensitive inferences have to be regulated as tightly as personal data, because they are being used as personal data – even when they are wrong. Citizens should be asked for their consent whenever external signs are being used to infer private information about them. They should have the power to contest and rectify inaccurate inferences, and inferred sensitive information should be treated as personal data.

With the abolition of microtargeted ads, personal data trades, default personal data collection, and sensitive inferences, privacy is already looking much better. But these measures are still not enough, because we still need to take care of the contexts in which personal data is not sold but may be used against the interests of citizens.

IMPLEMENT FIDUCIARY DUTIES

In most countries, the law does not force criminal suspects to self-incriminate. There is something perverse about making people complicit in their own downfall. A federal judge in

California banned police from forcing suspects to swipe open their phones because it is analogous to self-incrimination.[36] And yet we tolerate innocent netizens being forced to give up their personal data, which is then used in all sorts of ways contrary to their interests. We should protect netizens at least as much as we protect criminal suspects. Our personal data should not be used as a weapon against our best interests.

To accomplish such an objective, we should bind institutions that collect and manage personal data to strict fiduciary duties.[37] Fiduciaries such as financial advisers, doctors, and lawyers owe a duty of loyalty and care to their customers – and so should companies that hold personal data.

The word *fiduciary* comes from the Latin verb *fidere*, to trust. Trust is at the heart of fiduciary relationships. First, because the fiduciary is entrusted with something very valuable – your finances, your body, your legal affairs, or your personal data. Second, because by entrusting this valuable good to others, you are made extremely vulnerable to them. By accepting what is entrusted to them and in acknowledgement of your vulnerability, fiduciaries owe you their trustworthiness.[38]

Fiduciary duties exist to protect individuals who are in a position of weakness against professionals who are supposed to serve them but who might have conflicting interests. Your financial adviser could make excessive trades from your account to earn more commissions, or she could use your money to buy securities for herself. Your doctor could perform a surgery on you that is too risky or unnecessary simply as an opportunity to practise his skills, or to add a data point to his research. Your lawyer could sell your secrets to another client whose interests oppose yours. And, as we have seen, those who collect your data can give it to data vultures, criminals, and so on. None of these professionals should abuse the power that has been given to them by virtue of their profession.

Fiduciary duties, then, are appropriate when there is an economic relation in which there is an asymmetry in power and knowledge, and in which a professional or a company can have interests that go against the interests of their customers. Financial advisers, doctors, lawyers, and data experts know much more about finance, medicine, law, and data, respectively, than we do. They might also know more about you than you know yourself. Your financial adviser is likely to have a better grasp of your financial risks. Your doctor understands what is happening in your body better than you do. Your lawyer will have a deeper understanding of your legal case. And those who analyse your data may know (or may think they know) much more about your habits and psychology than you do. Such knowledge should never be used against you.

Fiduciaries must act in the best interests of their customers, and when conflicts arise, they must put their customers' interests above their own. People who do not want to have fiduciary duties should not accept being entrusted with valuable personal information or assets. If you don't want to have the duty to act in your patients' best interests, then don't become a doctor. Having a desire to perform medical interventions on people's bodies is not enough. The job comes with certain ethical expectations. In the same way, if companies do not want to have to deal with fiduciary data duties, they should not be in the business of collecting personal data. Wanting to analyse personal data for research or commercial purposes is all very well, but such a privilege comes with responsibilities.

Critics of the idea that fiduciary duties should apply to big tech have pointed out that such a policy would go against tech companies' fiduciary duties towards their stockholders. According to the law in Delaware – where Facebook, Google, and Twitter are incorporated – directors have to 'treat stockholder welfare as the only end, considering other interests only to the extent that doing so is rationally related to stockholder welfare'.[39]

That companies should only work towards the benefit of their stockholders to the detriment of their customers seems like a morally dubious policy – particularly if the business in question has negative effects on the lives of millions of citizens. Morally, the economic interests of stockholders cannot possibly trump the privacy rights and democratic interests of big tech's billions of users. One option to work around this problem is to establish that whenever stockholders' interests come into conflict with users' interests, fiduciary duties to users take priority. Another option is to institute such large fines for breaches of fiduciary duties towards users that it's in the stockholders' best interests for companies to honour those duties, if they care about their bottom line.

Fiduciary duties would go a long way to ensure that the interests of big tech are aligned with the interests of their users. If tech companies want to risk our data, they should risk their business in the process. As long as tech companies can risk our

data and be sure that we are the only ones who will pay the
bill – through exposure, identity theft, extortion, unfair discrimination, and more – they will continue to be reckless.

With fiduciary duties added to the list, the data landscape is
greatly improved. Our data no longer gets shared or sold or
used against our interests. However, our personal data could
still be lost through negligence, which is why we need to implement higher cybersecurity standards.

IMPROVE CYBERSECURITY STANDARDS

Our privacy will not be adequately protected so long as the apps,
websites, and gadgets we interact with are insecure. Data is too
easy to steal. As things stand, companies have little motivation to
invest in cybersecurity. Cybersecurity is expensive, and it is not
something users appreciate, because it is invisible. Netizens don't
have an easy way of comparing security standards between
products.[40] We know roughly what a safe door looks like, but
there are no comparable tell-tale signs on apps or websites.

Not only do businesses not have much to gain from investing in cybersecurity, they don't have enough to lose when
things go wrong. If data gets stolen, customers are the ones
who bear most of the brunt. If a company is deemed to have
been grossly negligent, it might be fined, but if the fine is not
big enough (e.g. if it is lower than what it would have cost to
invest in cybersecurity), companies will be tempted to consider such fines as a tolerable cost of doing business.

Cybersecurity is a collective action problem. Society would be better off if everyone had acceptable cybersecurity standards. Institutions' secrets would be better protected, and they could enjoy their customers' trust. Citizens' data would be safe. And national security would be likewise safeguarded. But it is not in the interest of most companies to invest in security because it gives them few advantages, and it's expensive, which can put them at a disadvantage with respect to their competitors. In the current situation, insecure products can drive secure products out of the market, as there is no return on investing in cybersecurity.

Government regulation is how security is improved. If it weren't for governments implementing standards, things like buildings, drugs, food products, cars, and airplanes would be much less safe. Companies very often complain when they are first required to improve their security standards. Car companies famously resisted obligatory seatbelts. They thought they were ugly, and that users would hate them. In fact, users were quite happy to be safer. Over time, companies come to embrace regulation that protects them and their customers from security disasters. And they come to appreciate that regulation is sometimes the only way a company can invest in something valuable that does not have an immediate return without incurring a competitive disadvantage, because everyone else has to do it too.

Even though much of the privacy we have lost since 2001 has been the direct or indirect consequence of governments supposedly prioritizing security, experience has taught us that security and privacy are not a zero-sum game. When we erode our privacy, more often than not we undermine our security. The internet was made unsafe to allow corporations and governments to snatch our data so they could, in theory, keep us safe. The reality is that an unsafe internet is extremely dangerous for individuals, companies, and societies.

If our gadgets are insecure, hostile regimes can spy on our government officials. Rogue operators could take down the power grid of a whole country by hacking a few tens of thousands of power-hungry devices like water heaters and air conditioners and causing a bump in the electricity demand.[41] They could also take control of nuclear power plants,[42] or even nuclear weapons.[43] A massive cyberattack could shut down an entire country.[44] It is one of the two most prominent catastrophic threats that governments around the world have identified in their risk registers. The other one is a pandemic.

For *decades* experts have been warning about the risk of pandemics. Not only have societies continued to engage in the risky practices we know cause them (e.g. wet markets and factory farming), we have not prepared for them either. The coronavirus pandemic caught us without enough protective equipment for health professionals, for instance – something that is unforgivable, given what we knew. Human beings are able to prevent what they haven't lived through before, but it's not easy. Using our imaginations to foresee what can go wrong is vital to motivate us into action.

Imagine being in lockdown and your country suffering a massive cyberattack. The internet crashes. Maybe the electricity is down too. Even your landline, if you still have one, might be down. You can't reach your family, you can't call your doctor, you can't even access the news. You can't go out because of the pandemic. It gets dark early and you have one candle left to burn (who keeps boxes of candles any more?). Your electric heating doesn't work. You don't know what has happened, and you don't know when or if normality will resume.

This scenario is not so far-fetched. After all, cyberattacks spiked as a result of the coronavirus pandemic.* With so many people working from home with insecure Wi-Fi and unsafe devices, the 'attack surface', or possible points of entry, increased. Britain's electricity system's administrator was hit by a cyberattack during lockdown; fortunately it didn't affect the electricity supplies.[45] Attacks against the World Health Organization increased fivefold during the same period.[46] It's only a matter of time before a massive cyberattack happens. We know this, just as we knew a pandemic would happen sooner or later. We have to be better prepared, and we have to take action now if we are to have the slightest chance of preventing or mitigating such an event.

To improve our cybersecurity, it is crucial that we disconnect systems.[47] There is a current trend towards making everything connect to each other: your speakers to your phone, your phone to your computer, your computer to your television, and so on. If tech enthusiasts had it their way, the next point of connection

---

* A group of dodgy characters sit around a table. One of them has a gun. 'For health and safety reasons, we'll be transitioning to cyber crime' reads this recent cartoon by Paul Noth published in the *New Yorker*.

would be your brain. It's a bad idea. We use fire doors to contain possible fires in our homes and buildings, and watertight compartments to limit possible flooding in ships. We need to create analogous separations in cyberspace. Every new connection in a system is a point of possible entry. If all your devices are connected, it means that hackers can potentially gain access to your phone (a relatively sophisticated, sensitive, and secure device, if you have a good one) through your smart kettle (most likely an insecure system). If all our national systems are likewise connected, a cyberattack could bring them all down through just one system.

Initially, enhanced cybersecurity standards will mostly be about patching up bad systems. But eventually, security has to be baked into the way technology is designed. At the moment, for example, authentication in the connection protocols between your smartphone and the cell towers it connects to is inadequate. Your smartphone gives up sensitive data to every tower. That is why IMSI-catchers can vacuum up your data, as we saw in Chapter One.[48] We have to start designing all technology with hackers in mind. The time when the internet could resemble country houses with no fences, doors, or locks ended years ago. We have to catch up with reality.

DELETE DATA

With no more personal ads, no data vultures, no default data collection, fiduciary duties, and strong cybersecurity, our privacy landscape has been largely rebuilt. But what about all our personal data that's already out there, and the personal data that will be justifiably collected in the future? We need to delete the personal data that has been collected surreptitiously and illegitimately. Even in the case of personal data collected in legitimate ways for necessary purposes, there should always be a plan to delete data. With few exceptions (such as birth records), no personal data should be collected without a plan or the ability to delete it.

In his book *Delete*, Viktor Mayer-Schönberger argues that forgetting is a virtue we should recover in the digital age. The ability to forget is an important component of a healthy life. Just imagine not being able to forget anything that you have experienced. Researchers studied the case of Jill Price, a woman in California who lacks the gift of forgetting her experiences. She was able, for example, instantly to recall what she'd done every Easter from 1980 to 2008. Just like that, without prior warning or preparation. Her memory was so rich that it overshadowed her present. It had made her neither happy nor particularly successful in her career. She is a relatively ordinary person who feels anxious and lonely in the company of her overcrowded memories.

Cognitive psychologist Gary Marcus has hypothesized that Price's extraordinary memory may not be the result of having an unusual brain, but rather of an obsessive-compulsive disorder that doesn't allow her to let go of the past.[49] Having permanent records by default might be recreating that kind of obsessiveness in us all, or at least some of its negative characteristics.

People who remember too much wish they could turn off (sometimes, at least) an ability that can end up feeling like a curse. When your mind latches on to the past, it is hard to move on, to put both tragic and happier times behind you and live in the present. It is hard to accept what you have in front of you if the pull of better and worse times feels just as vivid. Worse times can make you sad, and better ones can inspire nostalgia. Constantly remembering everything others have said and done can also make people overly resentful.

Forgetting is not only a virtue for individuals, but also for societies. Social forgetting provides second chances. Expunging old criminal records of minor or juvenile crimes, forgetting bankruptcies, and erasing the records of paid debts offer a second chance to people who have made mistakes. Societies that remember it all tend to be unforgiving.

We have never remembered as much as we remember today, both as individuals and societies. Before computers came along we had two ways of forgetting: voluntarily, by burning or shredding our records, and involuntarily, by not being able to record most events and naturally forgetting them, or by losing our records through accidents and wear and tear.

For most of history, keeping records has been hard and pricey. Paper used to be extremely expensive, and we needed a fair amount of space to store it. Writing demanded time and dedication. Such constraints forced us to choose what we wanted to remember. Only a tiny fraction of experience could be preserved, and even then memory was shorter-lived than it is now. Back when paper was not acid-free, for instance, it disintegrated rather quickly. Such documents had an inbuilt expiry date set by the materials they were made of.[50]

The digital age has turned the economics of memory on its head. Today, it is easier and cheaper to remember it all than to forget. According to Mayer-Schönberger, four technological elements have contributed to turning remembering into the default: digitization, cheap storage, easy retrieval, and global reach. Experiences are automatically transformed into computer data, which is stored in increasingly smaller storage devices that are increasingly cheaper. We then retrieve our data at the request of a few keystrokes, and send it anywhere across the globe with a click.

Once data collection became automated, and storage became so cheap that it was suddenly realistic to aspire to collect it all, we went from having to select what to remember to having to select what to forget. Since selecting takes effort, forgetting has become more expensive than remembering by default.

It is tempting to think that having more data will necessarily make us smarter, or able to make better decisions. In fact, it may impede our thinking and decision-making capabilities. Human forgetting is partly an active process of filtering what is important. Not selecting what we remember means that every piece of data is given the same weight, which makes it harder to identify what is relevant in a sea of irrelevant data.[51]

We are collecting so much data that it is impossible for us to get a clear picture from it – our minds have not evolved to cope with such vast amounts of information. When we have too much data and we're trying to make sense of it, we face two options. The first is to select a bit of information based on some criterion of our choosing that might make us blind to context in a way that can reduce our understanding, rather than increase it. Imagine you have a fight with a friend over Brexit. Ruminating over your fight, you decide to re-read all your text messages that contain the word 'Brexit'. These messages might be ones that do not exemplify your relationship as a whole, they only show a disagreement, but brooding over them might lead you to end your friendship. Had you remembered all the good times you'd had that were not captured digitally, or had you read the messages in which your friend supported you through a rough patch, you would have been reminded of why you were friends.

The second and increasingly common option to try to make sense out of inordinate amounts of data is to rely on algorithms as filters that can help us weave a narrative, even though they have no common sense to know what is important. For instance, an algorithm designed to determine who is a criminal from analysing facial images might end up picking out people who are not smiling. The algorithm doesn't have the necessary reasoning capacity to understand that, in its training data, the images of criminals provided by the police were ID photos in which people were not smiling.[52] Furthermore, algorithms have been shown time and again to suffer from biases embedded in our data, in the assumptions we make about what we are trying to measure, and in our programming. I recently came across someone who claimed to trust algorithms more than human beings because people make too many mistakes. How easily we lose sight of the fact that it is people who create algorithms, and often technology does not only not correct our mistakes – it amplifies them.

Handling too much data, then, can lead to less knowledge and worse decision-making. The double risks of twisting the truth and of memory being an obstacle to change combine to make permanent and extensive records about people positively dangerous. Such records capture people at their worst and freeze them in that image, not allowing them to fully overcome their mistakes. Old personal data can also lead us to biases tied to our history: if we use old data to determine the future, we will be prone to repeating the mistakes of our past.

We need to introduce expiry dates and forgetting into the digital world. We could design tech so that whatever data gets created self-destructs after a period of time. Some apps are already doing it: you can set an expiry date on your text messages on Signal, for instance. We could do the same with the files on our computers, our emails, our searches online, our purchasing histories, our Tweets, and most other data trails.

Whatever technological means we choose to use, the bottom line is that the default should not be to hold on to personal data indefinitely. It's too dangerous. We need to have methods that allow for periodic deletion of personal data that is no longer needed.

A critic might argue that one cannot ethically force a society to forget. Democracies do not distinguish themselves by forcing oblivion. Burning books and deleting online posts are marks of authoritarian governments, not democracies. The natural tendency for stable societies respectful of citizens' rights is to accumulate data, so the argument goes. Such reasoning would be convincing if we didn't have the capability of retaining data for ever. There is nothing natural about permanent records. Nature used to impose oblivion on us through the capacity of forgetting, and now that we have defied that natural process, we are realizing the price is too high. We have to reintroduce what is natural in a context – the digital world – that is as far from nature as it gets. Importantly, data should never be deleted on ideological grounds. A government shouldn't get to delete data that makes it look bad. Only *personal* data should be deleted, and only on the grounds of respecting citizens' rights, without discriminating with regard to its political content.

There is something to be said, however, for holding on to certain kinds of data. Much of what we have learned from history, for instance, comes from personal diaries. Some data we should delete entirely, but in some cases, a minority, perhaps it is enough to put locks on data to make it less accessible, or accessible only under certain circumstances (e.g. after the person has died, or a hundred years from the time it was created). We can bequeath some of our data to our loved ones, and in particular, our children and grandchildren, so they can trace back their roots. With the consent of the relevant people, perhaps we can keep a small fraction of personal data under strong safeguards that can be representative of a certain time and place for historians to learn from in the future.

The locks for that data should not only be legal (as laws change and are broken) but also technical (e.g. using encryption), and practical. By practical I mean making it difficult for people to access that data. If a diary is kept in paper format in a registry in one town, it is accessible to serious researchers, but it makes it harder for rogue actors to get to it than if it were published online and indexed on search engines. The extent to which something is accessible matters. That is the essence of Europe's right to be forgotten.

When Mario Costeja did a Google Search on his name in 2009, some of the first items to come up were a couple of notices from the late 1990s in the Spanish newspaper *La Vanguardia*. The notices were about Costeja's house being auctioned to recover his social security debts. They had first been published in the paper edition of the newspaper, which was later digitized.

Costeja went to the Spanish Data Protection Agency to complain against *La Vanguardia*. He argued those notices were no longer relevant, since his debts had been settled. Having that stain linked to his name was hurting his personal and professional life. The newspaper had refused to delete the records, and the Spanish Data Protection Agency agreed with it – *La Vanguardia* had published those public records lawfully. But the agency did ask Google to delete the link to the auction notice. A person who has paid his debts should not be burdened with that weight for the rest of his life.

Google appealed the decision, and the case ended up in the European Court of Justice, which in 2014 ruled in favour of the right to be forgotten. Costeja's records can still be found in *La Vanguardia*, but are no longer indexed in Google Search. Although the implementation of this right has given rise to doubts and criticism, its principle makes sense. It is questionable to have private companies be the arbiters of whether a request to make something less accessible has merit, even if the decision can be appealed and forwarded to a Data Protection Agency. What matters most, however, is that the right to be forgotten protects us from being haunted by personal data that is 'outdated, inaccurate, inadequate, irrelevant, or devoid of purpose, and when there is no public interest'.[53]

Unless we relearn how to forget in the age of the machine, we will be stuck in the past, both as societies and as individuals. It won't always be easy, however, to make sure our data has been deleted, or if it hasn't been deleted, to oversee how it's being used. We have no access to institutions' databases, which is why we may need to develop ways of tracking our personal data.

TRACKING OUR PERSONAL DATA

One of the biggest challenges of regulating personal data is the difficulty of policing data. At the moment, we are forced to rely on the word of tech companies that have proved themselves to be untrustworthy. Data protection authorities in Europe are often understaffed and underfunded. It is hard to have regulatory bodies that can oversee all institutions dealing with personal data. Tech giants are currently more powerful and wealthy than many countries. Limiting the use of personal data in the ways I suggest will make the job of policing easier, if only because there will be less personal data sloshing around. But personal data will still be hard to monitor.

That individuals have no way of knowing who holds their data is a huge handicap. It worsens already worrisome asymmetries between individuals and institutions, and it puts the burden of identifying abuse solely on supervisory agencies.

It would be ideal if we could track our own data. Imagine having an app that could show you a real-time map of who has your data and how they are using it, and allowed you instantly to withdraw your personal data if you wanted to. One of the most frightening aspects of the digital age is that, as you read these words, it is possible you are being subjected to dozens of algorithms that are judging your personal data and deciding your fate on the basis of it, all without your knowledge or consent. Right this moment, an algorithm may be tagging you as not being creditworthy, while another algorithm is deciding to put you further down on the waiting list for that surgery you need (possibly on the basis of a misguided criterion), and yet another is considering you unemployable. If you don't know when an algorithm sifts through your data and makes a decision about you, how can you ever realize that you might be the victim of an injustice? If you can't track who has your data and how they are using it, how can you be sure your data rights are being respected?

There are at least two major technical challenges associated with enabling people to track their data. The first is to match personal data to the person whom the data is about, and make sure that everyone is asked for their consent before data about them is shared. There are easy cases: we need only your consent to collect or use your email address. But when personal data includes data about more than one individual, it can get complicated. To share your genetic data ethically, you need consent from your parents, siblings, and children, at a minimum. But what if your children are still underage and would not consent if they were adults? And what about your cousins? How far down the family line should we ask for consent? It is hard to answer these questions because we cannot be sure about what kinds of inferences might be made about your cousins with your genetic data in the future. When in doubt, we should err on the side of caution. Perhaps we should not allow people to share their genetic data except with their doctors when there is a serious medical need.

The second major challenge is to devise a way in which we can be thoroughly informed about how our data is being used without further endangering our privacy. It's a tall order, and it may not be possible. It may be that by further tagging personal data in order to track it, we inevitably expose ourselves, in which case the main objective – to better protect our privacy – would be ultimately undermined. It remains to be seen. The creator of the World Wide Web, Sir Tim Berners-Lee, is currently working on a project, Solid, that aims to develop personal data pods that give users complete control over their personal information. If Solid or a similar project manages to overcome these and other technical problems, it could dramatically change the way we manage personal data.

CURB GOVERNMENT SURVEILLANCE

Governments do not need to engage in mass surveillance to keep citizens safe. Data collection and analysis should not happen without an individual warrant (as opposed to a bulk warrant), and it should only happen if it's necessary. It should also be targeted (as opposed to mass surveillance), and proportionate to the circumstance. According to the *New York Times*, at least 2,000 law enforcement agencies in all 50 states in the United States have tools to break into locked, encrypted phones and download their data. And sometimes law enforcement searches phones for minor crimes, such as an investigation into a fight over $70 at a McDonald's. That is not proportionate use of extremely invasive surveillance.[54]

Governments should not ask companies to build in backdoors for them, and they should not subvert cybersecurity. It is more important for national security that all citizens have safe devices, than that the government have access to all devices, because if your government has that access, so do other, possibly nefarious, actors.

There should be strong oversight of surveillance powers. Oversight bodies should have access to all relevant information. There should be an adequate amount of transparency to guarantee that citizens know the rules of their country. Individuals who are being surveilled should be notified either before it happens or, if that would jeopardize a criminal investigation, after the fact. People who have been surveilled should have access to their data, and be given the opportunity to correct it or provide relevant context.

Government-on-government espionage should be separated from surveillance. Government surveillance is the job of the military and the Foreign Office or State Department. Surveillance of private citizens is justified in criminal investigations and is the job of the police. The rules of espionage can be secret. The rules of surveillance should be known. Both spying and surveillance should be targeted.[55]

Whistleblowers should be protected from retaliation by law. Whistleblowers are the moral canaries of the coal mine that is society. Canaries are more sensitive than human beings to toxic gases such as carbon monoxide, which is why miners used them to detect danger. In some ways, whistleblowers are more sensitive than most others to injustice. They alert us of dangers to society. Once the canary showed signs of carbon monoxide poisoning, the miners would revive it with an oxygen tank. We have to make sure we have oxygen tanks for our whistleblowers.

Metadata is data about data. It is information that computers need to operate, and it is the by-product of such operations. Most of it cannot be encrypted, because if it were, computers wouldn't be able to communicate with each other. That makes it a hard problem for privacy. To limit the surveillance on metadata, it might be a good idea to spread internet traffic so that not too much of it passes through a few data centres, and we should use more onion routing (a technique to preserve anonymity) to protect that data.

Metadata includes information such as the operating system that created the data, the time and date of creation, the author of the data, and the location where the data was created. Metadata is more sensitive than it appears. You can infer things like someone having an abortion, for instance. 'If you have enough metadata, you don't really need content,' said former NSA General Counsel Stewart Baker. 'We kill people based on metadata,' said General Michael Hayden, former director of the NSA and CIA.[56] We should try our best to stop authoritarian regimes from getting access to metadata through infrastructure.

BAN SURVEILLANCE EQUIPMENT

There are some surveillance technologies that are so dangerous, so prone to abuse, that it might be better to ban them altogether, just as we ban some weapons that are too cruel and hazardous. We should consider banning facial recognition, as well as gait and heartbeat recognition and other technologies that destroy anonymity – they are ideal tools for oppression.[57] So are IMSI-catchers, software designed to break into smartphones, and other similar spyware; they should be illegal.[58] According to a lawsuit, software designed by Israeli surveillance company NSO Group has been used to hack the phones of activists, lawyers, journalists, and academics.[59] In addition to identification technologies and spyware, high-resolution satellites and drones are a third type of surveillance we should avoid.[60]

The Satellite Sentinel Project aimed to use satellite images to provide an early warning system to deter mass atrocities in Sudan. Two days after the project published satellite images of a new road in Sudan that they thought might be used to transport weapons, a rebel group ambushed a construction crew near an intersection in one of the photos and took twenty-nine people hostage. The timing of the publication of the images and the attack suggests they might be related.[61] Publishing high-resolution satellite images can be dangerous. Will Marshall and his team at Planet Labs already image the whole planet every single day through satellites. Now they're trying to index all the objects in the world and make them searchable. *Anyone* could surveil you using their satellites. That they are allowed to do this work, and that they present it with no mention of privacy, is very alarming.[62] The sky should not be watching us.

There is a race to map the public and private world, to create a digital duplicate of our physical world. Can you imagine how powerful an authoritarian regime could be if it had a detailed real-time map of every room and building in the world (including furniture and people living in those places)? If you have the latest Roomba vacuum cleaner, it is probably creating a floor plan of where you live.[63] Amazon has recently announced an autonomous indoor drone that can map your home and better surveil your family.[64] According to the promotional video of Facebook's Project Aria, the platform is already allowing 'a few hundred workers' to wear their smart glasses 'on campuses and in public spaces'. The video mentions advantages to having the world mapped, like being able to find your keys easily.[65] Surrendering our privacy is not worth it. There are other ways to find your keys that don't require creating a real-time virtual copy of reality. What right has Facebook – a private company with one of the worst privacy records – to create a duplicate of our world so they can better surveil us? None. That kind of data should not belong to a corporation. There are aspects and corners of reality that shouldn't even be turned into data in the first place.

FUND PRIVACY

In addition to investing in developing privacy tools, we need better data governance. Regulatory bodies in charge of data protection have to be better funded and better staffed. If data protection authorities are to stand a chance against tech titans, we have to give them enough tools. Some of the suggestions here are already part of laws like the GDPR and the California Consumer Privacy Act. But, thus far, data protection agencies in Europe have been overwhelmed, their resources unequal to the task of fulfilling their responsibilities.[66] First we need to put our full weight behind privacy agencies to make sure the law is enforced. And then we must regulate the data economy into oblivion.

CREATE DIGITAL REGULATORY AGENCIES

The European Union, Australia, and the United Kingdom are already discussing the creation of a big tech regulator. It makes sense to do the same in the United States. If there is the Securities and Exchange Commission for markets, the Federal Aviation Administration for airlines, the Food and Drug Administration for pharmaceuticals, and the Federal Communications Commission for telecommunications, big tech might well merit having their own specialized regulatory agency. [67]

UPDATE ANTITRUST LAWS

Antitrust regulation has to reflect the realities of power in the digital age. If a company can set abusive terms of service without losing customers, they should be investigated, whether they charge their users or not.

It is possible that big tech is so powerful, that we might need to break it up before we can regulate personal data properly. Pessimists tend to think that big tech is too big to ever be broken up or regulated, but this view lacks historical perspective. We have regulated every other industry that preceded it. Why would tech be any different? The United States' Justice Department filing an antitrust lawsuit against Google this year might be the beginning of seriously tackling big tech. A legitimate worry, however, is that antitrust suits can take a long time (as an example, Microsoft's case lasted eight years), and big tech moves very fast. That is why we must use many approaches to regulate big tech, not just one. A reason for optimism is that there are many countries in the world wanting to regulate tech. Having a common objective can give regulatory powers more clout, if nations around the world collaborate and coordinate.

DEVELOP DATA DIPLOMACY

Just like climate change, privacy is a collective action problem, and international agreements will be important in protecting

personal data. Data flows rarely respect borders.[68] If countries behave too individualistically, we might see the emergence of 'data
havens'. Similar to tax havens, a data haven would be a country
involved in 'data laundry', being willing to host data acquired in
unlawful ways that is then recycled into apparently respectable
products. That data could also be used for the purposes of training spyware that gets sold on to whoever is willing to pay for it,
including authoritarian regimes.[69] International pressure will be
paramount in improving privacy standards everywhere.

Another area in which we need diplomatic work is in the
kind of data that should be allowed to be shared between allied
intelligence agencies. We might come up with good rules to
curb state surveillance in our own country, but that effort will
be futile if our government can acquire the personal data of its
citizens from another country (e.g. the United States and the
United Kingdom often share data between them). Democracies are often left not knowing what data their government has
on them because their government claims they don't have the
authority to reveal data that was collected by an allied intelligence agency. We need clear rules and more transparency
regarding what data our intelligence agencies can request from
other countries.

PROTECT OUR CHILDREN

We should protect everyone, but children above all, because
they are in an exceptionally vulnerable position. Young

children depend on their families and schools to protect their privacy. And the current trend is to monitor them from the time they are conceived with the excuse of keeping them safe.

There are two fundamental reasons to worry about children's privacy in particular. First, surveillance can jeopardize their future. We don't want our children's opportunities endangered by institutions judging them (and likely misjudging them) on account of data about their health, their intellectual capacities, or their behaviour at school or with their friends. Equally, and possibly even more importantly, too much surveillance can break people's spirit. To bring up children under surveillance is to bring up subjects, not citizens. And we want citizens. For their own wellbeing and for the sake of society.

Society needs autonomous and engaged citizens able to question and transform the status quo. Great countries are not made up of servile followers. To develop into people with strong hearts and minds, children need to explore the world, make mistakes, and learn from their experiences, knowing that their blunders will not be recorded, much less used against them. Privacy is necessary to cultivate fearlessness.

Probably on account of their extreme vulnerability, children, and in particular teenagers, are generally more sensitive than adults regarding what others think about them. Surveillance can therefore be all the more oppressive for them. Youths who are watched all the time will be less likely to dare to try something new, something at which they may be bad in the beginning but which they could well master with practice over time, if only they were left alone to make a fool of themselves without an audience.

What makes the case of children a difficult one is that they do need some amount of supervision to keep them safe. The risk is that safety will be used as an excuse for undue surveillance, and the line between what is necessary and what is unwarranted is not always obvious.

Defenders of school surveillance argue that they are 'educating' students in how to be good 'digital citizens', getting them used to the ubiquitous surveillance they will be exposed to after they graduate. 'Take an adult in the workforce. You can't type anything you want in your work email: it's being looked at,' said Bill McCullough, a spokesperson for Gaggle, an American company that monitors schools. 'We're preparing kids to become successful adults.'[70] No. What excessive surveillance does is teach kids that human rights do not have to be respected. We cannot realistically expect people who have been taught as children that their rights do not matter to have any respect for rights as adults.

Surveilling children from the outset so that they get used to it for their adult lives is like implementing a completely unfair system of grading in school so that kids get used to the fact of unfairness in life. If we don't accept the latter, we shouldn't accept the former. A fair grading system not only gives all children the same access to opportunities, it also teaches them to expect fairness from institutions, which later in life will encourage them to demand, fight for, and create fairness when it is not offered. The same applies to privacy.

Surveillance teaches self-censorship. It is a warning for students not to push limits, not to talk about or even search online for sensitive topics; any behaviour outside the limits of what is politically and socially safe could trigger a school investigation, or even a police inquiry. But teenage years are all about curiosity about the stuff of life. Youths wonder about sex, drugs, and death, among other sensitive issues, and discouraging them from exploring those topics will not contribute to their knowledge or maturity. In a sense, when we supervise young people too heavily, we curtail their becoming responsible adults who will not need supervision. By oversurveilling children, by oppressing them with a Thought Police, we risk bringing up a generation of people who were never allowed to grow up.

As parents, there is much you can do to protect your children's privacy. But before we move on to what we can do as individuals to protect privacy, there is a common objection that is worth responding to.

DON'T WE NEED PERSONAL DATA?

Enthusiasts of the data economy will surely tell you that pulling the plug on the torrent of personal data will hamper innovation. The most alarmist version of this claim argues that if we regulate the data economy, foreign, possibly adversarial, powers will develop AI faster than we will, and we will be left behind. To limit what we can do with data is to put a stop to progress, so this argument goes.

The short answer to this objection is 'no' – progress is defending people's human rights, not undermining them. The benefits of the data economy, in terms of profits, scientific advancement, and security, have been consistently exaggerated and its costs have been downplayed by tech enthusiasts.

There is a longer answer. Even if we take progress to mean only technological progress, the answer is still 'no' – protecting privacy does not have to come at the cost of advancing technology. Let's not forget that much of what personal data is used for is financial gain. Businesses like Google may not even have to innovate to have a sustainable business model. Maybe Google's chances of directly selling its services to consumers were not great at its inception. Consumers hadn't had enough of a taste of what it is like to navigate daily life with the help of Google's search engine, Maps, and other products. But now netizens have a good sense of the value of Google's products. Let us pay for those services if we value them enough. In 2013, Google was already doing extremely well. It had about 1.3 billion users, and it had a yearly revenue of about $13 billion. It earned about $10 per year from every user.[71] Wouldn't that be a reasonable price to pay for Google's services? It's less than what people pay for entertainment services like Netflix, which costs a bit more than $10 per month.

Allowing personal data to be profitable creates an incentive to collect more of it than is necessary for technological progress. Using data for the purposes of science and tech will still be allowed, but if institutions want to experiment with *personal* data, they will need to take on the appropriate responsibilities to respect people's rights. It's a reasonable ask. If tech companies manage to transform their efforts into valuable services for netizens, then we will be happy to pay for them, just like we are happy to pay for other things we value in the offline world.

Furthermore, it is far from clear that infinite troves of personal data will necessarily result in technological and scientific progress. As we've seen, too much data can impede thinking and decision-making. Feeding more data into a bad algorithm will not make it a good one. What we are striving at when we design artificial intelligence is to build, well, just that – intelligence. If you have interacted with a digital assistant in recent times you will have noticed that they are not very bright.

Human beings can sometimes learn new things with one example, and they can transfer that knowledge to similar but new scenarios. As AI systems become smarter, we can expect them to need less data.[72] The most important challenges to the development of AI are technical ones, and they won't be solved by throwing more data at the problem.[73] In light of these observations, it should not surprise us that arguably the most sophisticated contributions of AI thus far have not been brought about through the exploitation of personal data.

AlphaZero is an algorithm developed by Google's DeepMind that plays the ancient Chinese game of Go (as well as chess, and shogi). What makes Go a particularly interesting game for AI to master is, first, its complexity. Compared to chess, Go has a larger board, and many more alternatives to consider per move.

The number of possible moves in a given position is about 20 in chess; in Go, it's about 200. The number of possible configurations of the board is more than the number of atoms in the universe. Second, Go is a game in which intuition is believed to play a big role. When professionals get asked why they played a particular move, they will often respond with something to the effect that 'it felt right'. It is this intuitive quality that makes people consider Go an art, and Go players artists. For a computer program to beat human Go players, then, it would have to mimic human intuition – or, more precisely, mimic the results of human intuition.

What is most remarkable about AlphaZero is that it was trained *exclusively* through playing against itself. It used no external data. AlphaGo, the algorithm that preceded Alpha-Zero, was partly trained through being shown hundreds of thousands of Go games between human beings. It took Deep-Mind months to train AlphaGo until it was able to beat the world champion, Lee Sedol. AlphaZero developed super-human Go abilities in three days. With no personal data.

WHAT ABOUT MEDICINE?

Medicine constitutes a very special case in the world of data. First, because medicine is extremely important to all of us. We all want to live longer and healthier lives. Everyone wants medicine to advance as fast as possible. Second, because medical data is very sensitive – it can lead to stigmatization, discrimination, and worse. Third, because it is tremendously hard and sometimes impossible to anonymize medical data. Genetic data, as we have seen, is a good example: it is the data that identifies you as *you* – it incorporates your very identity. More generally, for medical data to be useful, it is important to identify different data points belonging to one and the same person, and the more data points we have on someone, the easier it becomes to identify them.

Does the advancement of medicine depend on *trading* personal data? No. First, we should be more sceptical about the power of digital tech. Second, there are ways to use personal data for the purposes of medical research that minimize risk to patients and that ask for their consent. Third, some of the most important medical advances might not need to use personal data at all. Let's take a closer look at these three points.

Medical digital tech in perspective

Digital tech and big data are not magic. We can't expect them to solve all of our problems. Sometimes the innovations that save more lives are not those of high-tech, but rather less glamorous changes, like better hygiene practices. This is not to say that high-tech cannot contribute to medicine, but we shouldn't leave our critical minds outside the door whenever we discuss tech. When tech becomes an ideology, as sometimes it does, it veers away from science towards superstition. Here are two examples of how digital tech has overpromised and underperformed in the context of medicine.

The first case is IBM's artificial intelligence, Watson. In 2011, after Watson had managed to defeat two human champions in *Jeopardy!*, an American quiz-style game show, IBM announced that its AI would become a doctor. The company said that it expected its first commercial products to be available in eighteen to twenty-four months. Nine years later, that promise is yet to be fulfilled.

In 2014, IBM invested $1 billion in Watson. By 2016, it had acquired four health data companies for a total of $4 billion. Yet many of the hospitals that entered into projects with IBM Watson have had to end them. The MD Anderson Cancer Center had to cancel its project with Watson to develop an advisory tool for oncologists after having spent $62 million on it.[74] In Germany, the university hospital of Giessen and Marburg also gave up. When a doctor told Watson that a patient was suffering from chest pain, the system didn't consider that he might be having a heart attack. Instead, it suggested a rare infectious disease.[75] On another occasion, Watson suggested that a drug be given to a cancer patient with severe bleeding that could have caused the bleeding to worsen. 'This product is a piece of s—,' concluded a doctor at Jupiter Hospital in Florida.[76]

Watson is not an isolated case of tech disappointing in medicine. In 2016, DeepMind made a deal with the Royal Free NHS Trust in London. DeepMind got the medical records of 1.6 million patients, without their consent or knowledge. That means the company got access to pathology reports, radiology exams, HIV status, details of drug overdoses, who'd had an abortion, who'd had cancer – all of it.[77] The Information Commissioner's

Office would later deem that Royal Free breached data protection laws.[78]

The original idea was to use AI to develop an app to detect acute kidney injuries. Researchers soon realized that they didn't have good enough data to use AI, so they settled for something simpler. In the end, Streams, the app that was developed, has been shown to have 'no statistically significant beneficial impact on patients' clinical outcomes'.[79]

These two failures do not imply that all attempts will fail, but they give some perspective to the promises of digital tech in medicine. A recent meta-analysis looked at around 20,000 studies of medical AI systems that claimed to show they could diagnose illness as well as doctors. Researchers found that only fourteen of those studies (less than 0.1 per cent) were of sufficient methodological quality to try those algorithms in a clinical setting.[80]

That medical AI might not help patients is not the only concern. A more important worry is that it might hurt patients. For instance, AI might lead to overtreatment. Some medical digital technologies seem to err on the side of false positives (detecting medical problems when there are none). Certain algorithms looking for cancerous cells, for instance, will label as anomalous perfectly healthy cells at a rate of eight false-positive mistakes per image.[81] If companies and doctors have an interest (financial, professional, or data-driven) in intervening with patients, that might lead to a problem of overtreatment.

Yet another possible problem is glitches. It is dangerous to depend on digital tech because programming is extremely hard, and digital tech has needs that analogue tech doesn't have, all of which leads to digital tech often being less robust a technology than analogue. Compare an e-book to a paper book. The e-book reading device needs to charge every certain amount of time, it can get hacked, it relies on an internet connection, it can get ruined if you drop it into sand or water or on to a hard surface, and so on. In contrast, paper books are remarkably hardy. No need to charge, and if you drop it from the top of a building it'll probably survive (though the person walking by might not, so don't). When we're dealing with life-saving equipment, we want technology as robust as a paper book.

These failures are only meant to provide some realism and perspective on the potential of digital tech in medicine. Of course, AI might turn out to play a very important role in advancing medicine. But, as with any other intervention, we need it to be based on hard evidence before we are asked to hand over our personal data, and we need some reassurances that our data will be handled properly, and that the benefits are shared in a fair and equitable way. Too often AI gets a free pass.

Suppose we decide that we do want to do medical research with personal data and digital tech. After all, the promises of personalized medicine are attractive. There are ways to do it that are more ethical than how DeepMind and Royal Free did it.

Ethical medical research

Medical ethics has a long history of recruiting people for research. Research with personal data should not be seen to be too different from other kinds of medical research. Even though donating personal data feels nothing like donating blood – no needle, no pain – there are likewise risks involved. We don't force people to sign up for clinical research any more (though we used to, before medical ethics came along). We shouldn't force people to sign up for medical research with their personal data either. It is not acceptable to use the general population as guinea pigs without their consent, without proper safeguards and without compensation. Rather, we should ask for their consent, set some rules for how data will be used and when it will be deleted, and compensate research subjects appropriately, just like we do for other kinds of research.

Sometimes, public health institutions do not have the kind of resources or technology to analyse data, and they might benefit from collaborating with industry. In such cases we should make absolutely sure that the deals struck are beneficial to data subjects and patients. Among the many mistakes Royal Free made, two stand out as particularly egregious. First, they didn't secure any legal guarantee that DeepMind wouldn't use that data for anything other than developing the app. They got a promise that they wouldn't pair the data with data held by Google, but when DeepMind's health division got absorbed by Google, privacy experts feared that promise would be broken.[82]

The second big mistake was that Royal Free did not make sure that patients would benefit from products developed with their data.[83] Public health institutions have so much precious medical data that they have negotiation clout – they should use it. They should limit companies' access to that data. Perhaps companies could use the data but not keep it, for instance. And they should ask for legal guarantees that any product developed will be offered to public health institutions and the public at affordable prices.

It will always be a struggle to keep personal data safe when interacting with companies that do not have the public good as their main objective. But if we are lucky, perhaps the most important medical advances that AI can offer will not be the product of working with personal data at all.

Medical advances without personal data

As we saw, AlphaZero, AI's poster child, is an extraordinary feat, but it does not have practical applications in daily life (not yet, anyway). One way in which AI might change (and possibly save) our lives is through discovering new drugs.

Antibiotics are quite likely the most important medical advance of the last century. Before antibiotics, the leading causes of death worldwide were bacterial infectious diseases. Most people in developed countries now die much later than they otherwise would have, and from non-communicable diseases such as heart conditions and cancer.* Unfortunately, the effectiveness of these wonder drugs is now under threat on account of antibiotic resistance. Through evolutionary processes such as mutations, bacteria are growing resistant to the antibiotics they

---

* At the time of writing, it is still unclear whether the coronavirus pandemic will claim enough lives to count as an exception to most people dying from non-communicable diseases.

have been exposed to. The more we use antibiotics, the more opportunities these bacteria have to cultivate resistance. A world without effective antibiotics is a realistic and very scary prospect. Surgical procedures that are considered low risk would become high risk. Many more women would die after childbirth. A trip to the dentist or a one-night stand that results in an infection could kill you. Chemotherapy and organ transplantation would be much more dangerous, as these treatments depress the immune system. Antibiotic resistance could be a major contributing factor to a fall in life expectancy.

We desperately need new antibiotics, but the process of discovering and developing new drugs is slow and expensive. Researchers at MIT, however, think they might have developed a way to find new antibiotics. Through feeding a computer program information on the atomic and molecular characteristics of thousands of drugs and natural compounds, they trained an algorithm to identify the sorts of molecules that kill bacteria. They then gave the algorithm a database of 6,000 compounds. The algorithm selected one molecule predicted to have strong antibacterial power, and that, importantly, has a different chemical structure from existing antibiotics.

This computer model can screen more than 100 million chemical compounds in just a few days, something that would be impossible to do in a regular lab.[84] The hope is that the new antibiotic, halicin, will be strong and will work in new ways to which bacteria have yet to develop resistance. Similar advances could be possible in the discovery of antiviral and antifungal drugs, as well as vaccines. If AI helps us win the arms race against the superbugs, it will have earned its place in medicine.

That two of the most outstanding advances in AI have been made without any kind of personal data may not be a coincidence: personal data is frequently inaccurate, and can become obsolete relatively quickly.

The upshot is that we will not be hampering the development of AI by protecting our privacy. We can use personal data, with the appropriate precautions, but we don't have to make it into a commodity. And we may not need it for most advancements. The mark of true progress is the protection of citizens' rights and the enhancement of people's wellbeing. On both those counts, the personal data trade has nothing to show for itself.

BEWARE OF CRISES

As I write this chapter, the coronavirus pandemic is still raging. Tech and telecom companies across the world have offered their data collection and analysis services to governments to try to stop the contagion. Google and Apple agreed to join forces to modify their software to support the development of contact-tracing apps.[85] It is a dangerous time for privacy. When there is panic in the air, there is a tendency to be more willing to renounce civil liberties in exchange for the feeling of safety. But are coronavirus apps going to make us safer? It is far from clear.

In the town of Vò, where the first coronavirus death in Italy was recorded, the University of Padua carried out a study. Researchers tested all the inhabitants. They discovered that infected but asymptomatic people play a fundamental role in the spread of the disease. They found sixty-six positive cases whom they isolated for fourteen days. After two weeks, six cases continued to test positive for the virus. They had to remain isolated. After that, there were no new cases. The infection was completely under control. No app needed.[86]

Contact-tracing apps are bound to be less precise than tests, telling some people to stay at home even if they are not infected (despite having been near someone infected) and allowing others to roam freely who are infected and should be isolated. Apps cannot replace tests because they work through proxies, as opposed to testing whether someone has the virus.

What we need to know is whether someone has contracted the coronavirus. What apps do is try to find ways to infer infection. But they all have problems, because what counts as a 'contact' for an app is not the same as getting infected. Apps often define a 'contact' as being close to a person (within two metres) for fifteen minutes or more. The first thing to notice is that apps work on phones. If you don't have your phone with you, the app doesn't work. But suppose everyone carries their phone around (maybe by law, which would be gravely intrusive). We can trace contacts through GPS or Bluetooth. Neither is perfect.

The app might identify two people being in contact who are in fact on different floors of the same building, or who are on the same floor but separated by a thin wall. If they are

contacted and warned about possible infection, they will constitute false positives. But apps can also be expected to result in
a high number of false negatives. Suppose you bump into a
friend in the street and it's been so long since you last saw each
other that you immediately hug and kiss without thinking
twice about it. If you are not Mediterranean or Latin, maybe
you shake hands. Either way, you may get infected and the app
wouldn't suspect you because you didn't spend fifteen minutes
together. Or you might get infected through a contaminated
surface. In both these cases, the app would not identify you as
someone at risk. In such cases, the app might create a false
sense of security that can make people act less carefully than
they otherwise would.

Tracking everyone with apps when in most countries only
hospitalized people or people with symptoms are being tested
for the virus makes little sense. The apps will notify people who
have been in contact with those who have tested positive for
coronavirus, but by that time, those people will have infected
others, who in turn will have infected others, many of them
remaining asymptomatic and spreading the infection further.

Most infected people will not end up in hospital, and many
will not get tested (under current policies at the time of writing
in most countries around the world). To contain the spread of
the virus, then, we need massive testing. And if we have massive testing, it's not clear that we would benefit from apps. If we
had access to coronavirus tests that were cheap and simple to
use so that every person could do a daily test at home, we
wouldn't need an app – we would already know who's got the

virus, who has to stay home and who can go out. Six months
into the pandemic, most countries still didn't have the testing
capacity that would allow us to stop the spread through identifying those who had become infected.

There is a dangerous narrative circulating that argues that
what allowed China to better control the pandemic than more
democratic countries was their authoritarianism; in particular,
their use of an extremely intrusive app. What is more likely is
that their system of massive testing was what did the heavy lift-
ing. In May 2020, China tested the entire city of Wuhan in ten
days. In October 2020, it tested all of the population of 9 million
people in the city of Qingdao in five days after a dozen cases of
infections were discovered.[87] No Western country has performed that kind of mass testing at scale.

In addition to being imprecise, any app will also introduce
privacy and security risks. The easiest way to get hacked is probably by turning on your Bluetooth. Using the app, tech-savvy
people could potentially learn who infected them or a loved
one – a dangerous piece of information for a disease that can be
fatal. Or someone could leverage the system to surveil app users,
or generate heatmaps of where infected people are, for instance.
Yves-Alexandre de Montjoye and his team have estimated that
trackers installed on the phones of 1 per cent of London's population could allow an attacker to track the real-time location of
over half the city.[88] Remember, privacy is collective.

Why, then, were apps prioritized, as opposed to thorough
testing of the population? Perhaps because they are cheaper.
Perhaps because tech companies are the main big players in

today's context. Whenever there is a crisis, companies are asked to help. Maybe if the world's biggest companies were manufacturers, they would be offering to produce hand sanitizer, masks, gloves, and ventilators. In contrast, what big tech has to offer are apps and surveillance. It's not that an app is exactly what is needed in this situation and we are coincidentally lucky to live in the age of the surveillance economy. It may be more a case of hammers looking for nails.

Perhaps apps were prioritized because there is often magical thinking surrounding tech – hope that it will miraculously solve all of our problems. Perhaps because there are economic incentives to collect data. Perhaps because governments were at a loss as to how to solve the crisis, and accepting the many offers of apps they received was an easy way to show the public that they were doing *something*. (That it might not be helpful and perhaps harmful was a different matter.) Perhaps it was a combination of things. But no app can be a substitute for our medical needs. What we need are medical tests to diagnose people, a good system to support people who need to self-isolate, protective equipment and vaccines to prevent the disease, and medicines and other resources to treat patients. Apps are not magic wands, and having more data and less privacy is not the solution to our every problem.

The 2020 pandemic was not the first emergency situation that put privacy at risk, and it won't be the last. We have to learn how to handle situations like these in a better way. 'Never let a serious crisis go to waste,' said Rahm Emanuel, Barack Obama's Chief of Staff; it is 'an opportunity to do things that you think you could not do before.[89] In her book *The Shock Doctrine*, Naomi Klein extensively documented instances in which disasters have been seized as opportunities to pass extreme policy initiatives that enhanced the powers of the state.[90] When crisis ensues, citizens are distracted, scared, and more at the mercy of their leaders. Too often, that ends up being a bad combination for democracy. Extraordinary circumstances are taken advantage of to impose new normals that would never have been tolerated by the citizenry in less exceptional times. And few changes are as durable as those that were meant to be temporary.

Remember, that's how we got here in the first place. We accepted extraordinary measures in the wake of 9/11, and those measures still haunt us. In China, events like the 2008 Beijing Olympics and the 2010 World Expo were used to introduce surveillance that remained in place after the events were over.[91] Many of the surveillance measures that were imposed to control the coronavirus were draconian, and citizens are right to fear their persistence. We should be extremely vigilant about how our data is being used. Already there has been abuses of track-and-trace data. In the United Kingdom, for instance, contact-tracing data collected at pubs and restaurants (and that was supposed to be used only for public health purposes) was sold on.[92]

Tech giants are jumping at the opportunity to extend their reach into our lives. It's not just about contact-tracing apps. Lockdown was used as a lab for a permanent and profitable no-touch future.[93] Digital surveillance is gaining ground in entertainment, work, education, and health. Tech billionaires like Eric Schmidt are pushing for 'unprecedented partnerships between government and industry'.[94] Palantir, the CIA-backed company that helped the NSA spy on the whole world, is now involved with both the United Kingdom's National Health Service[95] and the United States' Department of Health and Human Services, as well as the Centers for Disease Control and Prevention.[96] The NHS gave Palantir all kinds of data about patients, employees, and members of the public, from contact information to details of gender, race, work, physical and mental health conditions, political and religious affiliation, and past criminal offences.[97]

Some hope that the pandemic will result in a renewal of social welfare and solidarity, much like what happened after the Second World War. Except this time around, welfare is being tied to private companies and digital surveillance tools and platforms.[98] Holding welfare at ransom in exchange for data amounts to preying on the disadvantaged. In February 2020, a Dutch court ruled that welfare surveillance systems violate human rights; it ordered the immediate halt of a programme for detecting welfare fraud.[99] Other countries should take heed. Good governments cannot acquiesce in the systematic violation of citizens' fundamental right to live free from surveillance. Privacy should not be the price we have to pay to access any of our other rights – education, healthcare, and security prime among them.

The coronavirus killed many more New Yorkers than 9/11. Will we repeat the mistakes that we made then? One of the dangers of appealing to threats such as terrorism and epidemics to justify privacy invasions is that *those threats are never going away*. The risk of a terrorist attack or an epidemic is perpetual. As we have seen, mass surveillance doesn't seem to keep us safer from terrorism. The jury is still out on whether it might help us stay safe from epidemics. It is highly doubtful. But even if it did, at what price? You would be quite safe from terrorism and epidemics if you confined yourself to your basement for ever more – but would it be worth it? At what point is a small increase in safety worth the loss of civil liberties? And can't we find ways to increase our safety that do not violate our right to privacy? Banning factory farming and wet markets in which wild animals are sold alive might be much more effective at preventing epidemics – not to mention the potential benefit to animals' welfare.

During a crisis, it is easy to want to do whatever is needed to stop the catastrophe that is wreaking havoc. But, in addition to thinking about how to contain an imminent disaster, we also have to think about the world that will remain after the storm passes.[100] By definition, crises pass, but the policies we implement tend to stick. Stopping a problem now in a way that will invite an even graver mess in the future is no solution. Before surrendering our privacy in the midst of a crisis we should be absolutely sure that it is necessary, and that we have a way of regaining control of our rights once the emergency is over. Otherwise we might end up in a deeper hole than the one we are trying to escape.

THE TIME IS NOW

As powerful and inevitable as the tech giants seem, it is not yet too late to reform the data ecosystem. Many parts of the economy have not been digitized. In the West, before the coronavirus pandemic, only a tenth of retail sales happened online, and around a fifth of computing workloads were in the cloud.[101] The pandemic pushed us further into the digital realm. We have to be careful. If we allow tech giants to continue expanding without setting strict rules for what they can turn into data and what they can do with that data, soon it will be too late. The time to act is *now*.

WHAT YOU CAN DO

Most of the crucial social, economic, political, and technological changes societies have gone through at some point seemed inconceivable to a majority of the population. That goes for both positive and negative developments. Women's rights. Electricity. Liberal democracies. Airplanes. Communism. The Holocaust. The Chernobyl nuclear disaster. The internet. It all seemed impossible. And yet it happened.

The world can change rapidly and dramatically. At the beginning of March 2020, life's humdrum routine seemed steady enough. People coming and going, supermarkets fully stocked, hospitals functioning as normal. Within weeks, a third of humanity was in complete lockdown due to the coronavirus pandemic. Much international travel stopped, food shopping became a risky, sometimes challenging excursion, and health services were overwhelmed by the weight of cases.

Buddhist philosophy calls the changing nature of life 'impermanence'. The potential for transformation can be frightening, because it reminds us that things can get worse at any moment. But impermanence also allows things to improve – it means *we* can make things better. Things are bound to change, and it's up to us to harness that primal fact of life to do what we can to make sure change happens for the better.

The history of rights is, to a large extent, the history of progressively recognizing that human beings are not resources to exploit, but individuals to respect. Labour rights are particularly relevant in this context because there will always be some economic pressure to ignore them. We first recognized that all human beings have self-ownership rights, that it is unacceptable to treat people as property. Then we recognized that human beings are entitled to certain basic working conditions: a safe environment, decent working hours, appropriate pay, holidays, and so on.

That it might be profitable to do away with rights is beside the point: labour rights are human rights that constitute red lines. If capitalism is to be inhabitable, and compatible with democracy and justice, we have to limit it. We have to make sure that companies find ways of making a profit that do not entail destroying what we value the most.

In the past, social movements have been crucial to pass laws that recognize rights and improve society. Exploitative practices end when they are banned by law, but people like you and

me are the ones who have to change the culture so that laws
can be passed and enforced. To change our privacy landscape,
we have to write about it, persuade others to protect their and
our privacy, get organized, unveil the inner workings of the
abusive system that is the surveillance society, support alternatives, envision new possibilities, and refuse to cooperate in
our own surveillance.

Any social system depends on the cooperation of people.
When people stop cooperating, the system breaks apart. Often the
necessity of cooperation isn't obvious until it stops, and with it
the whole machinery grinds to a halt. The trade in personal data
relies on our cooperation. If we stop cooperating with surveillance capitalism, we can change it. If we look for privacy-friendly
alternatives, they will thrive.

In this chapter you can find advice on how better to protect
your privacy and that of others, from the very easy to the more
cumbersome. Not everyone will want to do everything they can
to protect privacy. Keeping your personal data safe at this stage of
the digital age can be inconvenient. How far you are willing to go
will depend on how strongly you feel that protecting your privacy
is something you ought to do, and what your personal circumstances are. If you are an activist working in an undemocratic
country, it is likely you will be willing to do much to protect your
privacy. If you live in a safe country, have a stable job, and are not
looking to ask for a mortgage any time soon, you might be less
strict. The choice is yours. But before being too lax about your
privacy, keep in mind the following three considerations.

First, convenience is overrated, even if it is enticing. Like pleasure, convenience is an important component of a good life. It promises us an easier life. If we didn't choose convenience every now and then our lives would be hopelessly uncomfortable and inefficient. But convenience is also dangerous. It leads us to have sedentary lifestyles, to eat junk food, to support companies that harm society, to have monotonous and unsatisfactory daily routines, to be uneducated and politically apathetic. Exercising, reading, learning, inventing new ways of living and interacting, and fighting for just causes are as inconvenient as they are meaningful. The most satisfying achievements in life are seldom the easiest. A good life demands a reasonable degree of struggle – the right balance between the ease of convenience and the benefits of meaningful effort. Like pleasure, convenience has to be weighed against the price we have to pay for it, and the consequences that are likely to ensue.[1]

Second, the choices you make today will determine how much privacy you enjoy in the future. Even if you think you have nothing to hide today, you might have something to hide a few years from now, and by then it might be too late – data that has been given away can often not be recalled. Your country might be respectful of your human rights today, but can you be absolutely sure that it will continue to be so in five or ten years' time?

Third, how much privacy you have influences the level of privacy of your loved ones, your acquaintances, your fellow citizens, and people who are like you. Privacy is collective and political – it's not just about you.

With these cautions in mind, here are some things you can do to better protect privacy.

THINK TWICE BEFORE SHARING

You are one of the biggest risks to your own privacy. Human beings are social beings, and many online platforms such as Facebook are purposefully designed to feel like our living room. But unlike in our (tech-free) living room, a myriad of corporations and governments are listening to us online. Next time you post something, ask yourself how it might be used against you. And let your imagination run wild, because sometimes it takes a good deal of ingenuity to envisage how your private information or photographs can be misused. For instance, most people think nothing of posting a photograph in which you can see parts of hands or fingers. But fingerprints can be read from photographs, and even cloned.[2] Bear in mind that photographs also contain metadata like location, time, and date. Do a search for how to delete this information from your photos before you upload them anywhere (the exact method varies from device to device). Generally, the less you share online, the better. Sometimes what you want to share is important enough to be worth the risk, but don't share unreflectively.

RESPECT OTHERS' PRIVACY

Privacy etiquette is important. Respect other people's rights. Before posting a photograph of someone else, ask for their consent. In turn, they are likely to ask you for permission next time they feel like posting something about you. In more naive times, most people thought it was enough to have the ability to 'untag' yourself from photos uploaded by others. Now we know that facial recognition can be used to identify you with or without a tag.

If someone takes your photograph or records you without your consent, don't hesitate to ask them not to post that content online. When I first started worrying about privacy, I was shy to make such requests. But the responses I've received since then have given me confidence that most people empathize with privacy concerns. To my surprise, most people not only are not annoyed or even indifferent to my privacy requests, but curious about my reasons and shocked that it had never occurred to them that it might be inconsiderate to share photographs of other people without first asking for their permission. As people are becoming savvier about the risks of sharing information online, it is becoming more common to ask for consent before posting anything on social media.

When you invite someone over to your place, warn them of any smart devices you have. Even Google's hardware chief Rick Osterloh recommended as much when asked at an event. 'Gosh, I haven't thought about this before in quite this way,' he said. That gives you a sense of how oblivious tech designers are to privacy and our welfare more generally. At least he was honest about it, and about admitting that owners of smart speakers should disclose them to their guests.[3] That's more than can be said for many tech bosses.

Your guests are not the only ones deserving of privacy. Keep in mind that children are also owed privacy. It is not right to upload pictures to social media of other people's children without their parents' permission. Not even if they're family.[4] And you should respect your own children's privacy too. Sonia Bokhari's parents kept her off social media until she turned thirteen. When she became old enough to join Twitter and Facebook, she realized her mother and sister had been sharing photographs and stories about her for years. She reported feeling 'utterly embarrassed, and deeply betrayed'.[5] Children are people too (even teenagers) – and people have a right to privacy. Don't share videos of children who are being ridiculed in any way, no matter how funny they might be. Those children might be bullied for it at school afterwards, and it might change the very way they see themselves. Be careful about uploading funny videos of your children, as they could become viral.

Don't do a DNA test for fun. They're wildly inaccurate anyway, and you'll be jeopardizing not only your own privacy, but also the genetic privacy of your parents, siblings, offspring, and countless other kin for generations to come.

Do not betray people's trust. Do not threaten to publish other people's private messages or photographs to get them to do what you want. That is called blackmail or extortion, and it is both illegal and immoral. Do not expose other people's private messages or photographs. Exposing other people when they have given you access to their private life is a betrayal, and it contributes to a culture of mistrust. Do not be complicit in exposure. If someone shows you something that exposes someone else's privacy, express your disagreement, and do not share it with others.

CREATE PRIVACY SPACES

The spaces in which we can enjoy privacy have shrunk. We need consciously to create privacy zones in order to claw back some areas in which creativity and freedom can take flight unimpeded. If you want to have a particularly intimate and cosy party, ask your guests not to take any photographs or videos, or not to post them online. If you want your students to be able to debate freely in class, set up some rules stating that participants are not allowed to record or post what goes on in the classroom. If you want to organize an academic conference that will encourage the exploration of controversial topics, or of work in progress, shut off the cameras and microphones. Ditch the phone when you're spending time with your family – leave it in another room, at least sometimes. There are some interactions that will never flourish under surveillance, and we will miss out if we don't allow space for them.

SAY 'NO'

Perhaps because we are social beings, we seem to be predisposed to say 'yes' to most minor requests others may have. When someone asks for your name, it doesn't feel like a big deal to give it to them, and it can feel quite antisocial to say, 'Sorry, no.' This tendency to say 'yes' gets exacerbated online, when we are asked for consent to collect our personal data.

The consent notice feels like an obstacle to what we set out to do – access a website – and the easiest way to get rid of the hurdle is to say 'yes'. It takes being mindful to resist the temptation, but it is worth it. Privacy losses are like ecological damage or health deterioration: no one act of littering, no one puff of a cigarette will bring about disaster, but the sum of them through time might. Every data point you give out or withhold makes a difference, even if it doesn't feel like it.

Some websites are particularly bad at accepting 'no' for an answer. Instead of having one button to reject collection of data by all their partners, they make you say 'no' to each one, one by one. If you reject cookies, those websites will not remember your answers, so you have to go through the process each time. It's annoying, and unfair, and if you get frustrated, just close that website and look for an alternative.

CHOOSE PRIVACY

There are many ways in which our privacy is being snatched away from us. The lack of privacy may sometimes feel inevitable, but it isn't always so. Although there are data collection practices that are almost impossible to avoid, we often have more options than those that are immediately obvious to us. And whenever we do have an option, it is important to choose the privacy-friendly alternative – not only to protect our personal data, but also to let governments and businesses know that we care about privacy. In what follows, I list some things to look out for to protect your privacy when buying or using products and services, and some alternatives to dominant and invasive products and services. The technology landscape changes so quickly that this list will likely not include the latest products, so you may want to do a quick search for new ones. The most important take-home message is not brand names, but rather what to look out for to better protect your privacy.

Devices

Choose 'dumb' devices over 'smart' ones whenever possible. A smart kettle is not necessarily an improvement over a good old-fashioned one, and it represents a privacy risk. Anything that can connect to the internet can be hacked. If you don't need to be heard or seen, choose products that do not have cameras or microphones.

Think twice before buying a digital assistant like Alexa or Google Home. By inviting microphones into your home, you may be destroying the very fabric of intimacy with your loved ones. If you already have one, you can disconnect it – they make great paperweights. If you decide to keep one of these spies, make sure to research the settings thoroughly and choose the most private options.

It is particularly important to choose wisely when buying laptops and smartphones. These devices have cameras and microphones, they connect to the internet, and they store some of your most personal information – all reasons to choose a trustworthy product. When choosing a brand, think about the country of origin and conflicts of interest that the makers of devices might have (e.g. if the phone manufacturer primarily earns money through exploiting personal data, buy a different phone).

It also helps to read the latest news about privacy. In 2018, the directors of the CIA, FBI, and NSA warned Americans against buying devices from Chinese companies Huawei and ZTE on account of the suspicion that such products contain government-controlled backdoors.[6] In 2019, a study of more than 82,000 pre-installed Android apps across more than 1,700 devices manufactured by 214 brands found these phones to be incredibly insecure.[7] Pre-installed apps are privileged software that may be difficult to eliminate if you are not an expert user and may be collecting and sending your data to third parties without your consent. Unless you are a techie who knows how to build privacy for your phone, it's probably a good idea to stay clear of Androids. And don't keep any apps you don't need – your phone's security is only as strong as your weakest app.

Messaging apps

The most important thing about messaging apps is that they offer end-to-end encryption, and that you trust the provider will not misuse your metadata, or will not store messages in the cloud insecurely. Even though WhatsApp provides such encryption, its being owned by Facebook introduces privacy risks. After Facebook acquired it, Brian Acton, one of the co-founders of the app, admitted, 'I sold my users' privacy.'[8]

The safest option, from the point of view of external threats, is probably Signal. One of my favourite features is the ability to set expiry dates on your messages – you can set them to disappear after they are seen. Telegram is also worth mentioning. Telegram has the advantage that when you delete a text, you can delete it from all phones, not just yours, at any time, which is a great feature protecting you from internal threats. Sometimes you realize that you shouldn't have texted something, or that you trusted someone who was not worthy of your confidence. The ability to recall our texts at will is something every messenger app should have. There are two huge disadvantages with Telegram, however. The first is that cryptographers tend to distrust its encryption – it is likely less secure than Signal.[9] Furthermore, conversations are not encrypted by default; you have to choose the 'secret chat' option. Both Signal and Telegram are free and easy to use. You'll be surprised by how many of your contacts already have one of these alternatives. For those who don't have it, just ask them to get it. Many people will be happy to have a safer messaging app.

Email

Emails are notoriously unsafe. An email might feel as private as a letter, but it is more like a postcard without an envelope. Avoid using your work email for purposes unrelated to work (and sometimes for work, too). Your employer can access your work email, and if you work for a public institution, your email may be subject to freedom of information requests. When choosing an email provider, look for privacy perks like easy encryption, and consider the country in which it is based. At the moment, the United States has looser legal restrictions on what companies can do with your data. Some options that might be worth looking into are ProtonMail (Switzerland), Tutanota (Germany), and Runbox (Norway). If you are patient and tech-savvy, you can use PGP (Pretty Good Privacy) to encrypt your emails.

Do not give out your email address to every company or person who asks for it. Remember, emails can contain trackers. If you get asked for your email address in a shop, it is usually possible to politely decline. If the shop assistant informs you that they need an email in order to sell you something, give them a fake one – they deserve it (more on obfuscation below). To make a point about it, I often say that my email is something like noneofyourbusiness@privacy.com.

If you are forced to share your email address because you have to receive an email in order to click on a link, try using an alternative address that contains as little personal information as possible to deal with untrustworthy parties. To escape as many trackers as possible, find the setting in your email provider that blocks all images by default. Another good technique is the 'email plus trick'. Suppose the email address you have opened to deal with commercial junk is myemail@email.com. When you are asked for your email by an annoying company, give them your email plus a name that can identify the company: myemail+annoyingcompany@ email.com. You'll still get the email, but you can block that address if the company becomes too annoying, and if the email is ever leaked, you'll know who's to blame.[10]

Search engines

Your internet searches contain some of the most sensitive information that can be collected about you. You search for things that you don't know about, for things that you want, for things that worry you. Since you search for what is going through your mind, your searches are a glimpse into your thoughts. Stop using Google as your main search engine. Change your default search engine on your browsers to one that does not collect unnecessary data about you. Privacy-friendly options include DuckDuckGo and Qwant. You can always go back to Google exceptionally if there is something you are having trouble finding, but in my experience that is becoming less and less necessary.

Browsers

If you want to limit the amount of information that can be linked to your profile, it is a good idea to use different browsers for different activities. Different browsers do not share cookies between them. (A cookie is that small piece of data sent by the websites you visit and stored in your computer by your browser.) Authentication cookies are used by websites to recognize you as a unique user when you come back to visit the page. Tracking cookies are often used to compile your browsing history so that advertisers will know what to show you. Choose one browser for websites for which you have to sign in, and another one to browse the web. Brave is a browser designed with privacy in mind. One of its many advantages is that it has a built-in ad and tracker blocker; it is also faster than other browsers. Vivaldi and Opera are also good options. So are Firefox and Safari, with the appropriate add-ons. Firefox has a feature, the Multi-Account Container, that isolates cookies according to containers you set up.[11] Sites in one container can't see anything from sites opened in another container. You have to be signed in with a Firefox account to use it though.

USE PRIVACY EXTENSIONS AND TOOLS

Privacy extensions can complement your browser. If your browser does not automatically block trackers and ads, you can use an extension to take care of that.

Adblockers are easy to find and install. About 47 per cent of netizens are blocking ads.[12] Once you enjoy the undisturbed peace that adblockers bring you, you will wonder how you ever put up with so many annoying ads jumping at you and distracting you for such a long time. Using adblockers also sends a clear message to companies and governments: we don't consent to this kind of advertising culture. If you want to be fair to companies that make an effort to show only respectful ads – contextual ads that respect your privacy and are not too jarring – you can disable your adblockers for those sites.

Privacy Badger, developed by the Electronic Frontier Foundation, can block tracking and spying ads. DuckDuckGo Privacy Essentials also blocks trackers, increases encryption protection, and offers a privacy rating from A to F that lets you know how

protected you are when you visit a website. In addition to protecting your privacy, blocking such invasive tools can speed up your
browsing. HTTPS Everywhere is another extension developed
by the Electronic Frontier Foundation that encrypts your communications with many major websites. You can find other
extensions that can automatically delete your cookies when you
close a tab, or clear your history after a certain number of days.

Keep in mind that there are untrustworthy extensions, however. Cambridge Analytica used extensions that seemed
harmless, such as calculators and calendars, to access a user's
Facebook session cookies, which allowed the firm to log into
Facebook as if they were that user.[13] Before using an extension,
do a quick search on it to make sure it's safe.

Think of the most private thing you ever do online. For that,
you might want to consider using Tor, a free and open-source
software that allows you to be anonymous online. Tor directs
internet traffic through a worldwide volunteer overlay network
of thousands of relays. When you request to access a website
through Tor, your request will not come from your IP address.
Rather, it will come from an exit node (analogous to someone
else passing the message along) on the Tor system. Such a laby-
rinth of relays makes it difficult to track which message
originates from which user. The advantages are that the websites you visit don't see your location, and your Internet Service
Provider doesn't see which websites you visit. The easiest way
to use this software is through the Tor Browser. The browser
isolates each website you visit so that third-party trackers and
ads can't follow you around.

There are a few disadvantages to using Tor. Because data goes through many relays before it reaches its destination, it makes your browsing sluggish. Some websites might not work as well. A further disadvantage is that you may attract the attention of intelligence agencies – but you may have already done that by reading this book, or any article about privacy online.[14] Welcome to the club. Although intelligence agencies may not be able to see what you're doing online when you're using Tor, they know you are using Tor. On the bright side, the more ordinary people use Tor, the less it will be deemed suspicious behaviour by authorities. Protecting your privacy is not illegal; it is outrageous that we are made to feel as if it is.

Virtual Private Networks (VPNs) are also a popular privacy tool. A good VPN can channel your internet traffic through an encrypted, secure, private network. VPNs are especially useful when you want to access the internet through a public network such as the Wi-Fi you can find at an airport or other public spaces. A public Wi-Fi network makes you vulnerable to who-ever set it up and to other people who are connected to it. Using a VPN protects you from everyone except the company behind it, which gets extensive access to your data. Make sure you can trust whoever is behind a VPN before using it. It is not easy to know who is trustworthy, but sometimes it is relatively obvious to know who isn't. It is not surprising, for instance, that Facebook used its VPN, Onavo Protect, to collect personal data.[15] As a general rule, if the VPN is free, you are probably the product, so stay away.

CHANGE YOUR SETTINGS

You should assume that all settings for all products and services are privacy-unfriendly by default. Make sure you change your settings to the level of privacy you aim to achieve. Block cookies on your browser, or on some of your browsers. It is especially important to block cross-site tracking cookies. If you choose more secure and private settings, it might impact the functionality of some sites. At least some of those sites will not be worth visiting. You can start with strict settings and modify as you go along, according to your needs. Consider using your browser on a private mode (though keep in mind that such incognito modes only delete traces of your online activity on your computer; they do not protect you from external tracking).

If you want to be extra cautious, check your settings once a year – companies change their terms and conditions all the time. The appropriate privacy settings are not always grouped in the same place, so finding them may not be as easy as it sounds. If you're struggling to find them, bear in mind that it's not you who is stupid, it's them abusing their power. It might be worth doing an online search for how to change your privacy settings for the usual suspects like Facebook and Google.[16] If you're lucky, you might find an app to do it for you (Jumbo claims to do just that for your Facebook account, and similar apps may be in the making).

DON'T CYBERHOARD

Getting rid of data that you don't need any more is the virtual equivalent of spring cleaning.[17] The less data you hoard, the less risk you accumulate. I admit that deleting data is hard. There is the nagging feeling that someday you might need some of that data, even if you haven't needed it in a decade. A sobering experience for me was losing much of the data on my phone a few years ago. At the time, it felt like a catastrophe. Looking back, I haven't missed that data at all. A less radical solution is to create a backup of the data you have online, storing it on an encrypted hard drive, and deleting it from the internet. Thanks to the GDPR, it has become easier to download your data from platforms, even if you're not European. For instance, it is easy to request to download your data from your account settings on Twitter, and then use an app to delete your old tweets.

Truly deleting digital information from your devices is sometimes a challenge because of the way computers work at present. When you delete a file from your computer, although it disappears from view, it is still there. The data hasn't been touched. Instead, the computer's map has changed. The computer pretends the file is no longer there and marks the space as free. That's why you can use recovery software. Someone with enough skills and motivation could find your deleted files. If you ever want to sell your laptop, for instance, make sure you

erase your files for real. The best way is to encrypt your hard
drive (which you should do anyway), and delete the key. That
makes it cryptographically inaccessible – encrypted data looks
like gibberish.[18]

CHOOSE STRONG PASSWORDS

Never ever use '123456', 'password', the name of your favourite
sports team or personal information like your name or birth
date for passwords. Avoid common passwords.[19] The most
important feature of a password is its length. Use long passwords, with lower and uppercase letters, special characters,
and numbers. Don't use the same password for all sites. Ideally,
don't use any password for more than one site. Consider using
a trustworthy password manager that can generate strong
passwords and save them for you. Consider using multifactor
authentication, but beware giving out your mobile number to
businesses that will use it for purposes other than your security. The ideal two-factor authentication is a physical key like
YubiKey.

USE OBFUSCATION

If a stranger stops you in the middle of the street and asks you
an invasive question, you can refuse to answer it and walk
away. The internet does not allow you to remain silent. It tracks

you and infers personal information about you, whether you want it to or not. Such intrusion is akin to someone asking for your number in a bar and refusing to take 'no, thank you' for an answer. If that person were to continue to harass you for your number, what would you do? Perhaps you would give them a fake number. That is the essence of obfuscation.

'Obfuscation is the deliberative addition of ambiguous, confusing, or misleading information to interfere with surveillance and data collection.'[20] In a context in which you are not allowed to remain silent, sometimes the only way to protect your privacy and express protest is to mislead. Of course, government institutions such as tax authorities have a justified claim to your personal information. But businesses do not always have a claim to your personal data. Consider giving companies to whom you do not owe your personal data a different name, birth date, email, city, etc. If you want to express protest through obfuscation, you can choose names and addresses related to privacy – myemailisprivate@privacy.com.

Sharing accounts or gadgets is yet another form of obfuscation. A group of teenagers in the United States were worried about tech giants, school administrators, college recruiters, and potential employers looking at their social media. They found a way to protect their privacy on Instagram – they share an account. Having a network of people sharing an account makes it harder for prying eyes to work out which activity belongs to whom.[21] Sharing devices is even better for privacy, as someone looking carefully at the data could infer which data belongs to whom based on their device, rather than their account.

GO ANALOGUE

Minimizing digital interactions is a good way to enhance privacy. Paper records kept under lock and key are probably safer than stored on your laptop. Pay with cash when possible instead of credit cards or smartphones. Go back to paper books; buy them at bricks-and-mortar shops. Leave your smartphone at home if you don't need it. When buying products, choose ones that do not connect to the internet. You don't need a kettle or a washing machine through which you can get hacked. Often, smart is dumb.[22]

BUY NEWSPAPERS

The free press is one of the pillars of free and open societies. We need good investigative journalism to tell us about what corporations and governments try to hide from us but shouldn't. If it weren't for the press, we might not know about the workings of surveillance capitalism. But for the press to work well, it needs to be independent, and if it is owned by power it risks serving power instead of serving citizens. We have to pay for the press so that it works for us. Buy (and read) newspapers. Keep well informed.

The digital age has been unkind to newspapers the world over. People getting 'free' content online were hesitant to pay for a newspaper subscription – never mind that free content was not really free (your data and attention were the price) and of questionable quality. The dominance of social media weakened the relationship between newspapers and their audience. People are increasingly getting news from social media. By accessing information through social media, you are more likely to be exposed to personalized content and fake news. Buy newspapers in paper, so that no one can track what you read. A second-best option is to visit newspapers' websites directly. Get your news from the source.

DEMAND PRIVACY

Demand that businesses and governments respect your data. Let's start with data brokers. There are too many of them to list them all here, but a few big ones are Acxiom, Experian, Equifax, and Quantcast. Privacy International has made this process much easier by providing templates and email addresses.* Another very useful tool to write and send data requests can be found in mydatadoneright.eu.

Let me warn you: it is an ordeal to email every company that has your data. Often they make it hard for you – they take a long time to respond, they ask you for more data (do not give it if it seems unreasonable), they may be evasive. Persevere as much as your patience and circumstances allow. Perhaps send those emails while you're queueing for the bus or at a

---

* https://privacyinternational.org/mydata

supermarket. And know that you might not be successful. But don't let that discourage you. Making the request is what matters most. It makes companies work (imagine if we were all to request our data), and it lets them know that the public does not consent to their practices. It creates a paper trail – evidence that policymakers can use to fine and regulate data vultures. At the very least, request your data for the companies that have more of your data, or more of your most sensitive data.

Demand privacy from every professional you interact with who asks for your data. Ask questions. Be careful with your medical data. Avoid using unnecessary health apps – they'll likely sell your data. Ask your doctor, dentist, and other health professionals about their privacy practices. Tell them that you don't consent to your data being shared in any way with anyone.

In order to demand privacy from companies and governments, it is important to know your rights. Read up on your laws. If you are a European citizen, know that you have a right to be informed, to access your data and rectify it, to ask for your data to be erased, to restrict the processing of your data, and to take your data to a different company, among other rights. If you have a complaint and have not been able to resolve a privacy issue with a company, you can contact your national data protection authority, or the European Data Protector Supervisor (depending on the nature of the complaint). Rights are worth little if they only exist on paper. We have to bring them to life.

Contact your democratic representatives. Send them an email, call them. Include them in your tweets about privacy. Tell them you are worried about your personal data. Ask them about their plans to protect your privacy. Vote for the right people. Politicians violating your right to privacy during their campaigns are a red flag – they do not deserve your vote.

When a company disappoints you with bad privacy policies, give them a bad review on websites like Trustpilot, and make sure you mention privacy in your complaint.

DON'T DEPEND ON THEM

Depending on any one tech company is dangerous. It means that part of your identity is in their hands, and if they cancel your account, or delete your emails (it happens), you may have much to lose. Tech companies want you to depend on them, so it is very hard not to. Sometimes it's impossible. But keep it in mind. There are degrees of dependency, and the less you depend on any one platform or app, the less power they'll have over you. Make sure you have your contacts in more than one place (preferably on paper), for example. Keep your personal connections alive in more than one way, so that at any moment you can close your account with any platform without too much loss.

ARE YOU IN TECH?

Maybe you work in one of the big tech companies we have been discussing. Maybe you work for a small start-up. Maybe you're designing your own app. Whatever the case may be, if you are part of the workforce constructing our digital architecture, you have a big role to play in baking privacy into your products from the start.

In addition to thinking about profit, those who build tech ought to ask themselves how they want to be remembered. As one of the people who helped companies and governments violate people's right to privacy, who put users' data at risk until something terrible happened? Do you want to be seen as one of the people who broke democracy? Or do you want to be remembered as one of the people who helped fix the data landscape by offering citizens a way to navigate life in the digital age while retaining their privacy?

One of the most chilling accounts of a tech company being on the wrong side of history is Edwin Black's *IBM and the Holocaust*.[23] It tells the story of how IBM contributed to the Nazi genocide through its punch card (see Chapter Four). The punch card was a powerful technology – it significantly enhanced the power of states to control people through categorizing and counting them – but it was nowhere near as powerful as the technologies that are being developed today. Facial recognition and big data inferences can facilitate a degree of control over people that is far more powerful to what we have known in the past. Reading that book made me hope that our grandchildren and great-grandchildren will not have to read a similar book on one of today's tech companies and a murderous regime of the future.[24] If tech companies want to be on the right side of history, they would do well to protect our privacy. In addition to privacy being a business opportunity, it is also a moral opportunity.

Companies and governments are made up of individuals, and while some individuals have more power than others to steer an institution in one direction or another, every individual is morally responsible for whatever they contribute to that institution. Programmers and tech designers are especially important in the digital age. They hold the expertise to make the machines do what we want them to do. They make the magic happen. Institutions covet computer scientists, engineers, and data analysts, which puts them in a good position to negotiate their responsibilities. If you work for tech and suspect you might be working on a project that could end up hurting people, you might want to consider pushing your employer towards more ethical projects, or even leaving your job and looking for work elsewhere (if you can afford to).

Tech workers can make more of a difference if they dissent together. In 2018, Google's workers managed to get the company to end forced arbitration of sexual harassment claims by employees, and not renew its contract for Project Maven, a collaboration with the Pentagon.[25] Dissenters can make a difference. Follow your conscience.

Alfred Nobel regretted inventing dynamite; Mikhail Kalashnikov wished he hadn't created the AK-47; Robert Propst came to hate what became of the office cubicles he designed; Ethan Zuckerman regrets inventing pop-up adverts. The list of inventors who come to repudiate their creations is a long one. Do not join that list. Good intentions are not enough; most inventors with regrets had good intentions. As an inventor you have to assume that someone will try to abuse whatever you create, and you have to make sure that cannot be done by design. It's a tall order.

People in tech can look to academics and not-for-profit organizations worried about privacy for advice. Following the work of people like Bruce Schneier, Cathy O'Neil (I recommend reading her *Weapons of Math Destruction*), and Yves-Alexandre de Montjoye, among others, might give you ideas. The Electronic Frontier Foundation, Privacy International, European Digital Rights and NOYB (from 'none of your business') are good sources of information. There are some ethics consultancies you can seek advice from; make sure they have a good reputation, and that there is someone trained in ethics involved (sounds basic, but it's not always the case). There are some organizations that help start-ups get off the ground which offer an assessment by an ethics committee as part of their programme.[26]

If you happen to be someone who is funding start-ups, make sure you require that the companies you fund undergo an ethical review of their products. Some start-ups will never worry about their ethics unless they are incentivized to noyb. They are too concerned about surviving and thriving, and they think privacy and ethics is something they can add on to the final product once they make it. Many bad things in tech happen simply because no one stopped to think how things might go wrong. Privacy and ethics have to be requirements from the very start of any tech project.

Tech designers and privacy-friendly companies can make a huge difference. Moxie Marlinspike and the secure messaging (and non-profit) app he created, Signal, have had an enormous impact on how we think about and use encryption.[27] Small new companies that offer privacy can snatch away business from mainstream companies, and big companies that up their privacy game can heavily pressure others to follow.[28]

DO YOUR BEST

Talk about privacy with your friends and family. Tweet about it. If you have a book club, read about privacy. In fiction, I recommend *Zed* by Joanna Kavenna, *The Circle* by Dave Eggers, and of course *1984* by George Orwell.

Turn off the Wi-Fi and Bluetooth signals in your smartphone when you leave home. Cover your cameras and microphones with a sticker. Take precautions when going through customs in countries that are known to be privacy-unfriendly.[29] Look out for opportunities to protect your privacy. And don't expect perfection.

All of these measures will make a difference. All of them can save you from violations of your right to privacy. But none of them is infallible. It is very hard to have flawless privacy practices. Even privacy experts frequently slip up. If you are tired, in a rush, or distracted, it is easy to give out more information than you want to. Furthermore, if someone is intent on invading your privacy, they will probably end up succeeding.

Even if you don't manage to protect your privacy perfectly, you should still try your best. First, you might succeed at keeping *some* personal data safe. That in itself might save you from a case of identity theft, or exposure. Second, you might succeed at keeping someone else's data safe, as privacy is a collective concern. Third, even if you fail at protecting privacy, such attempts have an important expressive function – they send out the right message. Demanding that institutions protect our privacy informs politicians and encourages policymakers to legislate for privacy. Choosing privacy-friendly products gives industry a chance to see privacy as a business opportunity, which will encourage them to innovate in our favour and to stop resisting regulation. Governments and companies are more worried than you might imagine about how you feel about privacy. We need to make it clear to them how much we care about our personal data.

You shouldn't need to do any of these things, and I hope your children will not need to take such precautions. Just as it is impossible for individuals to verify whether the ingredients in everything we ingest are edible – which is why we have regulatory bodies controlling that – it is impossible for individuals alone to solve the privacy problems we face. But it is up to us to motivate businesses and governments to protect our privacy. We can make it happen. And for our culture to start caring about privacy again, you don't need to attain perfection – doing your best is good enough.

REFUSE THE UNACCEPTABLE

I borrow this phrase from Stéphane Hessel's *The Power of Indignation*.[30] Hessel was a concentration camp survivor, a member of the French resistance, and was later involved in the drafting of the Universal Declaration of Human Rights. What do Stéphane Hessel, the abolitionists, Mahatma Gandhi, Martin Luther King, Rosa Parks, Nelson Mandela, Ruth Bader Ginsburg, and all the other heroes who have made the world a better place have in common? They refused the unacceptable. Our heroes are not people who inhabit injustices comfortably. They do not accept the world that has been given to them when it is an unacceptable world. They are people who dissent when it is necessary.

Aristotle argued that an important part of being virtuous is having emotions that are appropriate to the circumstances. When your right to privacy is violated, it is appropriate to feel moral indignation. It is not appropriate to feel indifference or resignation.

Do not submit to injustice. Do not think yourself powerless – you're not. At Microsoft's campus in Redmond near Seattle there is a room in which Azure, the firm's cloud-computing service, is managed. There are two big screens. One shows the status of the system. The other displays people's 'sentiment' about the system, as expressed on social media.[31] Why would a company like Microsoft care as much about how people feel about their system as they do about the functioning of the system itself? Because the latter depends on the former. The whole digital economy depends on you. On your cooperation and assent. Do not tolerate having your right to privacy violated.

The Universal Declaration of Human Rights is like a letter from those who came before us, warning us never to cross certain red lines. It was born out of the horror of war and genocide. It is a plea for us to avoid repeating the mistakes of the past. It warns that people will be 'compelled to have recourse, in the last result, to rebellion' if human rights are not respected. Privacy is a right for good reasons. Defend it.

CONCLUSION

What kind of society would you like to live in? Two worlds lie ahead. The first is a more extreme version of the surveillance society we live in today. It is a world in which your every step, word uttered, search online, purchase, and swipe of your finger on your smartphone is recorded, analysed, and shared with governments and companies. Drones and satellites watch you from above. Facial recognition identifies you wherever you go. Authorities track what you read, when you protest. Police, public health authorities, intelligence agencies, and surveillance companies receive this information. Your data is used mainly to prevent pandemics and terrorist attacks, authorities assure you. But you know it is also used for much more than that.

Surveillance is not only about what you *do*, but about what you *think* and *feel* – it's under-the-skin surveillance.[1] Your body is scrutinized to infer both your emotions and the state of your health. Your heart rate, temperature, and skin conductance (whether you are sweating) is assessed through your watch, which you may be forced to wear by law. Emotional surveillance companies record and analyse what makes you angry when you watch the news, what content online makes you afraid, and they share that data with the authorities.

They say such surveillance helps democracy. They say that you don't need to vote any longer because your government can infer what your political opinion is through data analytics. Your data allows the powerful to make predictions about your future on the basis of which decisions are made as to how you are treated in your society. Whether you get a job, a loan, or an organ donation if you need one is decided by surveillance and predictive algorithms.

This is a world in which machines manage you. They order the food you need to stay productive in the workforce when your fridge is running low. They time your efficiency at work, including your toilet breaks. They tell you to meditate when your stress levels increase. They tell you how many steps you have to take every day for exercise to keep your access to healthcare.

It is a world in which you worry about your children's privacy. You wonder whether their future might be compromised when they play games online, as you know their scores are being sold to data brokers that calculate cognitive capacities. You worry that they might make a mistake, such as getting drunk as teenagers and being photographed, and that they will never get a job as a result. You worry about how obedient they have to be to stand a chance in their society. You worry they will never taste freedom. This is a society primed for an authoritarian takeover.

But that is not the only future available to us. There is a better world awaiting. One in which what is yours is not exploited by governments and companies. One in which the data on your smartphone stays there, and no one has access to it except you. It is a world in which no one is allowed to share or sell your data, not even your family. It is a society in which you can go to the doctor and share your symptoms without worrying that this very act might harm you down the line. You can have a private conversation without it becoming public. You can make mistakes without them defining your future. You can search what worries you online, what you are curious about, without your interests coming back to haunt you. You can seek advice from a lawyer without suspecting that the government is listening in and fearing that you might be self-incriminating. You can rest assured that information about who you are, what you've gone through, what you hope and fear, and what you've done will not be used against you. This is a society in which the government's power derives from the consent of its citizens – not from their data. This is a society that continues and improves the millennia-long tradition of democracy.

A world in which privacy is respected is one in which you can go out to protest without fear of being identified. It's a world in which you can vote in secret. You can explore ideas in the safety of your mind and your home. You can make love without anyone except your partner tracking your heartbeat, without anyone listening in through your digital devices. You can enjoy intimacy – the kind that can only flourish between people who are alone together and who know that no one else is watching.

Not all tech is bad. A world in which we can enjoy privacy doesn't need to be one deprived of technology. We just need the right tech with the right rules in place. Good tech does not force-feed you. It is there to enhance your autonomy, to help you achieve your own goals, as opposed to tech's goals. Good tech tells it to you straight – no fine print, no under-the-table snatching of your data, no excuses, and no apologies. Good tech works for you. *You* are its client. Not advertisers, not data brokers, not governments. You're not only a user, and never a subject, but a citizen who is also a customer. Good tech respects our rights and our liberal democracies. Good tech protects your privacy.

Contrary to early claims that the digital age entailed the end of privacy, privacy is making its way back. This is not the end of privacy. Rather, it is the beginning of the end of surveillance capitalism. It will be a fierce battle, and one that we can never be complacent about winning once and for all. Rights have to be defended every day. Red lines that ought never to be crossed have to be repainted every season. It will require some time to take back control of our personal data. And we will need to do it together. But it can be done, and it will be done. The sooner, the better, to save ourselves many unnecessary risks and harms.

Six years ago, when I told people I was researching privacy, the most common response I got was bleak and cynical. 'Oh, so you're doing history, not philosophy.' 'Privacy is dead. Get used to it. Nothing to think about.' More sympathetic responses included attempts to ground my feet in reality, encouraging me to choose a topic of research with brighter prospects. In some ways, back then I was as pessimistic as the next person regarding privacy – the ferocity of the data economy was not leaving much space for hope. But I was also an optimist, in that I thought the nature and scale of the theft of personal data was so appalling and so dangerous that the situation was unsustainable – it just *had* to get better. I was right, and I am even more optimistic now. These days people respond to mentions of privacy with interest and concern.

The wind has changed. We are relearning the value of privacy after having temporarily forgotten it, dazzled as we were by the rise of digital technology. After the Cambridge Analytica scandal, and experiencing cases of public shaming or identity theft ourselves, we now understand that the consequences of today's lack of privacy are as severe as they were before the internet came along. The theft of your data can result in as expensive a bill as if your wallet was stolen. And data brokers knowing too much about you is even worse than when companies were able to ask you in a job interview whether you planned to have children. At least in the past they had to look you in the eye, and what they were doing was visible to all.

Politically, having our privacy compromised is more dangerous than ever. Never have we amassed so much personal data on citizens. And we have allowed surveillance to grow at a time when cybersecurity standards are poor, democracies are weak, and authoritarian regimes with a knack for hacking are on the rise. Digital tech used the cloak of data invisibility to erode our privacy. But now we know their tricks. We can take back control of our personal data.

The aftermath of the coronavirus pandemic is a major challenge to our privacy, but we are better off now than we were a few years ago. We know more about our privacy and how it is being exploited, there is more regulation on what institutions can do with our personal data, there are plans to further regulate personal data, and there is more pressure on tech companies to take privacy seriously. A few years ago no one thought that the GDPR was even possible. For all its faults, it's a major step in the right direction. And it's only the beginning.

We are currently witnessing a civilizing process similar to the one that made our offline life more bearable. Regulation made sure that food being sold was edible, that customers could return faulty products, that cars had safety belts, and that prospective employers couldn't legally ask you about whether you were planning to have children. The present historical moment is crucial if we want to tame the Wild West of the internet. The ground rules that we set now for personal data will determine the privacy landscape of the next few decades. It is critical that we get things right. We owe it to ourselves and to our children.

Privacy is too important to let it wither. Who you are and what you do is nobody's business. You are not a product to be turned into data and fed to predators for a price. You are not for sale. You are a citizen, and you are *owed* privacy. It's your *right*. Privacy is how we blind the system so that it treats us impartially and fairly. It's how we empower our citizenry. It's how we protect individuals, institutions, and societies from external pressures and abuse. It's how we mark out a space for ourselves in which we can freely relax, bond with others, explore new ideas, and make up our own minds.

It might seem radical to call for the end of the data economy. But it's not. It's just the status quo that makes it seem that way. What is extreme is having a business model that depends on the mass violation of rights. Widespread surveillance is incompatible with free, democratic, and liberal societies in which human rights are respected. It has to go. Do not settle for anything less. Data vultures will push back. Bad tech will apologize and say they'll do better while they ask for more of your personal data. Governments will team up with bad tech and promise you more safety in exchange for your data. Tech enthusiasts will tell you that progress will be hampered. But now we know better. Refuse the unacceptable. Take back control of your personal data, and privacy will prevail.

ACKNOWLEDGEMENTS

Taking back control of our personal data is a collaborative effort. This book is the result of innumerable acts of kindness by countless people, and the following list is bound to be incomplete. I hope those who are missing will forgive me, and still accept my warmest thanks for their generosity.

I wrote most of this book while in lockdown during the coronavirus pandemic. My sincere gratitude to all the key workers who risked their lives so that others could stay home.

I wish to acknowledge the Uehiro Centre for Practical Ethics, the Wellcome Centre for Ethics and Humanities, Christ Church, and the Faculty of Philosophy at the University of Oxford for their support over the past three years. Special thanks to Julian Savulescu, for making the Uehiro Centre a safe haven for cutting-edge research in applied ethics.

A writer could not ask for a better agent than Caroline Michel. Caroline was the first to see the potential in *Privacy Is Power* when the book was nothing but an idea – without her, the idea would have remained just that. My gratitude to all the team at Peters Fraser + Dunlop. Tim Binding was always encouraging and incredibly helpful in turning an incipient project into the beginnings of a book. Thanks also to Laurie Robertson, Rose Brown, and Rebecca Wearmouth.

I am thankful to everyone at Transworld. Susanna Wadeson and Patsy Irwin immediately shared my alarm about the state of our privacy, and set out to work with as much urgency as I exercised in writing this book. Stephanie Duncan was an exquisite editor to work with; thorough, patient, and solicitous. Thank you for going beyond the call of duty. Daniel Balado expertly copy-edited the final draft. Thanks to Katrina Whone, Vivien Thompson, Dan Prescott, and Cat Hillerton for producing such a beautiful finished book, Richard Ogle for producing the stylish jacket, and Sally Wray, Lilly Cox, and the sales and marketing teams for doing everything possible to help convey the messages within to as wide an audience as possible.

Thanks to Nigel Warburton, for inviting me to write a piece on privacy for *Aeon* that got me thinking about the relationship between privacy and power, and to *Aeon* (aeon.co) for allowing me to use that article as the basis for Chapter Three.

In the time I have been researching privacy, I have benefited from insightful conversations with many brilliant mentors and colleagues. My DPhil supervisors, Roger Crisp and Cécile

Fabre, have been crucial in testing and refining my ideas about privacy. Other important interlocutors include Anabelle Lever, Antonio Diéguez, Carina Prunkl, Ellen Judson, Evan Selinger, Gemma Galdon Clavell, Gina Neff, Gopal Sreenivasan, Katrien Devolder, James Williams, Jeff McMahan, Julia Powles, Julian Savulescu, Kevin Macnish, Lindsay Judson, Marjolein Lanzing, Peter Millican and all the speakers at the Ethics in AI Seminar at the University of Oxford, Tom Douglas, Václav Janeček, and Yves-Alexandre de Montjoye, among many others.

The following people very kindly read either parts of the book or the complete manuscript, and made it better with their feedback. Thanks to Bent Flyvbjerg, Javier de la Cueva, Ian Preston, Jo Wolff, Jorge Volpi, Diego Rubio, Yves-Alexandre de Montjoye, Mark Lewis, Marta Dunphy-Moriel, and Peter Millican for helping me hunt for mistakes. Any remaining errors are entirely my responsibility, of course.

I'm grateful to all my friends and loved ones, living near and far, for being there for me at every turn. Many thanks to Aitor Blanco, Alberto Giubilini, Areti Theofilopoulou, Daniela Torres, David Ewert, Diego Rubio, Hannah Maslen, Javier de la Cueva, Josh Shepherd, Kyo Ikeda, Luciano Espinosa, María Teresa López de la Vieja, Marina López-Solà, Marisol Gandía, Rafo Mejía, Ricardo Parellada, Rosana Triviño, Silvia Gandía, Sole Vidal, Stella Villarmea, Susan Greenfield, and Txetxu Ausín, among many others who have helped me throughout the years.

Gratitude doesn't even begin to convey what I owe my family. Héctor, María, Iván, Julián – no words would be enough. Heartfelt thanks to Ale and Alexis, and to the little ones. Lastly, thanks to Bent Flyvbjerg for always encouraging me to write, for reading next to me, for writing more than words with me. I will always be thankful that, despite the darkness and anxiety the pandemic brought, I was one of the lucky ones for whom lockdown had a silver lining: I was fortunate enough to be stuck at home with the right person, writing the right book.

NOTES

INTRODUCTION

1 Throughout the book I use 'the data economy', 'the surveillance economy', 'surveillance capitalism', and 'the surveillance society' almost interchangeably. We could, in theory, have a data economy that excludes *personal* data. We could trade in data that is about impersonal matters. But, at the time of writing, when people write about the 'data economy', they are often referring to the trade in personal data, so I use the 'data economy' as shorthand for the 'personal data economy'.

2 Remember how, in the first *Matrix* film, Trinity and Morpheus had to get to Neo through the Matrix to get him out of there?

3 Brittany Kaiser, *Targeted. My Inside Story of Cambridge Analytica and How Trump, Brexit and Facebook Broke Democracy* (Harper Collins, 2019), 81.

CHAPTER ONE

1 For more on self-tracking, see Gina Neff and Dawn Nafus, *Self-Tracking* (MIT Press, 2016).

2 Aliya Ram and Emma Boyde, 'People Love Fitness Trackers, But Should Employers Give Them Out?', *Financial Times*, 16 April 2018.

3 Ifeoma Ajunwa, Kate Crawford and Jason Schultz, 'Limitless Worker Surveillance', *California Law Review* 105, 2017, 766–767.

4 Sam Biddle, 'For Owners of Amazon's Ring Security Cameras, Strangers May Have Been Watching Too', *Intercept*, 10 January 2019.

5 Geoffrey Fowler, 'The Doorbells Have Eyes: The Privacy Battle Brewing Over Home Security Cameras', *Washington Post*, 31 January 2019.

6 Alex Hern, 'Smart Electricity Meters Can Be Dangerously Insecure, Warns Expert', *Guardian*, 29 December 2016.

7 Carissa Véliz and Philipp Grunewald, 'Protecting Data Privacy Is Key to a Smart Energy Future', *Nature Energy* 3, 2018.

8 L. Stanokvic, V. Stanokvic, J. Liao and C. Wilson, 'Measuring the Energy Intensity of Domestic Activities From Smart Meter Data', *Applied Energy* 183, 2016.

9 Alex Hern, 'UK Homes Vulnerable to "Staggering" Level of Corporate Surveillance', *Guardian*, 1 June 2018.

10 https://www.samsung.com/hk_en/info/privacy/smarttv/. Accessed 7 May 2020.

11 Nicole Nguyen, 'If You Have a Smart TV, Take a Closer Look at Your Privacy Settings', CNBC, 9 March 2017.

12 Matt Burgess, 'More Than 1,000 UK Schools Found To Be Monitoring Children With Surveillance Software', *Wired*, 8 November 2016.

13 Lily Hay Newman, 'How to Block the Ultrasonic Signals You Didn't Know Were Tracking You', *Wired*, 3 November 2016.

14 A version of this explanation was given by an Amazon spokesperson when Alexa recorded someone's private conversation and sent it to a random contact. Sam Wolfson, 'Amazon's Alexa Recorded Private Conversation and Sent it to Random Contact', *Guardian*, 24 May 2018.

15 Daniel J. Dubois, Roman Kolcun, Anna Maria Mandalari, Muhammad Talha Paracha, David Choffnes and Hamed Haddadi, 'When Speakers Are All Ears', *Proceedings on 20th Privacy Enhancing Technologies Symposium*, 2020.

16 Sam Wolfson, 'Amazon's Alexa Recorded Private Conversation and Sent it to Random Contact'.

17 Michael Baxter, 'Do Connected Cars Pose a Privacy Threat?', *GDPR: Report*, 1 August 2018.

18 Erin Biba, 'How Connected Car Tech Is Eroding Personal Privacy', BBC News, 9 August 2016; John R. Quain, 'Cars Suck Up Data About You. Where Does It All Go?', *New York Times*, 27 July 2017.

19 Bruce Schneier, *Data and Goliath* (London: W. W. Norton & Company, 2015), 68. IMSI stands for 'international mobile subscriber identity'.

20 https://www.aclu.org/issues/privacy-technology/surveillance-technologies/stingray-tracking-devices-whos-got-them. Accessed 15 October 2020.

21 Kim Zetter, 'How Cops Can Secretly Track Your Phone', *Intercept*, 31 July 2020.

22 Kim Zetter, 'How Cops Can Secretly Track Your Phone'.

23 Ben Bryant, 'VICE News Investigation Finds Signs of Secret Phone Surveillance Across London', *VICE*, 15 January 2016.

24 Losing your breath at the sight of your screen or email is a thing. It's called 'email apnea', or 'screen apnea'. Linda Stone, 'The Connected Life: From Email Apnea to Conscious Computing', *Huffington Post*, 7 May 2012.

25 Steven Englehardt, Jeffrey Han and Arvind Narayanan, 'I Never Signed Up For This! Privacy Implications of Email Tracking', *Proceedings on Privacy Enhancing Technologies* 1, 2018; Brian Merchant, 'How Email Open Tracking Quietly Took Over the Web', *Wired*, 11 December 2017.

26 Radhika Sanghani, 'Your Boss Can Read Your Personal Emails. Here's What You Need To Know', *Telegraph*, 14 January 2016.

27 Kristen V. Brown, 'What DNA Testing Companies' Terrifying Privacy Policies Actually Mean', *Gizmodo*, 18 October 2017.

28 Bradley Malin and Latanya Sweeney, 'Determining the Identifiability of DNA Database Entries', *Proceedings, Journal of the American Medical Informatics Association*, 2000.

29 S. Tandy-Connor, J. Guiltinan, K. Krempely, H. LaDuca, P. Reineke, S. Gutierrez, P. Gray and B. Tippin Davis, 'False-Positive Results Released by Direct-to-Consumer Genetic Tests Highlight the Importance of Clinical Confirmation Testing for Appropriate Patient Care', *Genetics in Medicine* 20, 2018.

30 Richie Koch, 'Using Zoom? Here Are the Privacy Issues You Need to Be Aware Of', (ProtonMail, 2020).

31 When communications are end-to-end encrypted, companies can't access their content, but Zoom could access video and audio from meetings, despite the desktop app stating that Zoom was using end-to-end encryption. Micah Lee and Yael Grauer, 'Zoom Meetings Aren't End-to-End Encrypted, Despite Misleading Marketing', *Intercept*, 31 March 2020. A few months later, Zoom fixed some of its privacy problems, but it announced it was going to exclude free calls from end-to-end encryption. After experiencing a privacy backlash, it then promised end-to-end encryption for all users. At the time of writing, months have passed and the company still hasn't started to roll out its end-to-end encryption. Kari Paul, 'Zoom to Exclude Free Calls from End-to-End Encryption to Allow FBI Cooperation', *Guardian*, 4 June 2020. Kari Paul, 'Zoom Will Provide End-to-End Encryption to All Users After Privacy Backlash', *Guardian*, 17 June 2020.

32 Michael Grothaus, 'Forget the New iPhones: Apple's Best Product Is Now Privacy', *Fast Company*, 13 September 2018.

33 Casey Johnston, 'Facebook Is Tracking Your "Self-Censorship"', *Wired*, 17 December 2013.

34 Kashmir Hill, 'How Facebook Outs Sex Workers', *Gizmodo*, 10 November 2017.

35 Kashmir Hill, 'Facebook Recommended That This Psychiatrist's Patients Friend Each Other', *Splinter News*, 29 August 2016.

36 Kashmir Hill, '"People You May Know": A Controversial Facebook Feature's 10-Year History', *Gizmodo*, 8 August 2018.

37 'Facebook Fined £500,000 for Cambridge Analytica scandal', BBC News, 25 October 2018.

38 Dan Tynan, 'Facebook Says 14m Accounts Had Personal Data Stolen in Recent Breach', *Guardian*, 12 October 2018.

39 Gabriel J. X. Dance, Michael LaForgia and Nicholas Confessore, 'As Facebook Raised a Privacy Wall, It Carved an Opening for Tech Giants', *New York Times*, 18 December 2018.

40 Kashmir Hill, 'Facebook Was Fully Aware That Tracking Who People Call and Text Is Creepy But Did It Anyway', *Gizmodo*, 12 May 2018.

41 Natasha Singer, 'Facebook's Push For Facial Recognition Prompts Privacy Alarms', *New York Times*, 9 July 2018.

42 Alex Hern, 'Facebook Faces Backlash Over Users' Safety Phone Numbers', *Guardian*, 4 March 2019.

43 Zack Whittaker, 'A Huge Database of Facebook Users' Phone Numbers Found Online', *TechCrunch*, 4 September 2019.

44 For a list of Facebook privacy disasters from 2006 to 2018, see Natasha Lomas, 'A Brief History of Facebook's Privacy Hostility Ahead of Zuckerberg's Testimony', *TechCrunch*, 10 April 2018.

45 Len Sherman, 'Zuckerberg's Broken Promises Show Facebook Is Not Your Friend', *Forbes*, 23 May 2018. 'Despite repeated promises to its billions of users worldwide that they could control how their personal information is shared, Facebook undermined consumers' choices,' said Federal Trade Commission Chairman Joe Simons. FTC Press Release, 'FTC imposes $5 billion penalty and sweeping new privacy restrictions on Facebook', 24 July 2019.

46 Allen St John, 'How Facebook Tracks You, Even When You're Not on Facebook', *Consumer Reports*, 11 April 2018.

47 Digital, Culture, Media and Sport Committee, 'Disinformation and "Fake News": Final Report' (House of Commons, 2019).

48 Brian Fung, 'How Stores Use Your Phone's WiFi to Track Your Shopping Habits', *Washington Post*, 19 October 2013.

49 Stephanie Clifford and Quentin Hardy, 'Attention, Shoppers: Store Is Tracking Your Cell', *New York Times*, 14 July 2013.

50 Chris Frey, 'Revealed: How Facial Recognition Has Invaded Shops – and Your Privacy', *Guardian*, 3 March 2016.

51 Kashmir Hill and Aaron Krolik, 'How Photos of Your Kids Are Powering Surveillance Technology', *New York Times*, 11 October 2019.

52 Yael Grauer, 'What Are "Data Brokers," and Why Are They Scooping Up Information About You?', *Motherboard*, 27 May 2018.

53 Adam Tanner, *Our Bodies, Our Data. How Companies Make Billions Selling Our Medical Records* (Beacon Press, 2017), 78, 95, 147–148.

54 Julia Powles and Hal Hodson, 'Google DeepMind and Healthcare in an Age of Algorithms', *Health and Technology* 7, 2017.

55 Daisuke Wakabayashi, 'Google and the University of Chicago Are Sued Over Data Sharing', *New York Times*, 26 June 2019.

56 Dan Munro, 'Data Breaches In Healthcare Totaled Over 112 Million Records in 2015', *Forbes*, 31 December 2015.

57 Alex Hern, 'Hackers Publish Private Photos From Cosmetic Surgery Clinic', *Guardian*, 31 May 2017.

58 Jennifer Valentino-DeVries, Natasha Singer, Michael H. Keller and Aaron Krolik, 'Your Apps Know Where You Were Last Night, and They're Not Keeping It Secret', *New York Times*, 10 December 2018.

59 Nick Statt, 'How AT&T's Plan to Become the New Facebook Could Be a Privacy Nightmare', *Verge*, 16 July 2018.

60 Joseph Cox, 'I Gave a Bounty Hunter $300. Then He Located Our Phone', *Motherboard*, 8 January 2019.

61 Olivia Solon, ' "Data Is a Fingerprint": Why You Aren't as Anonymous as You Think Online', *Guardian*, 13 July 2018.

62 Y. A. de Montjoye, C. A. Hidalgo, M. Verleysen and V. D. Blondel, 'Unique in the Crowd: The Privacy Bounds of Human Mobility', *Scientific Reports* 3, 2013.

63 Y. A. de Montjoye, L. Radaelli, V. K. Singh and A. S. Pentland, 'Identity and privacy. Unique in the Shopping Mall: On the Reidentifiability of Credit Card Metadata', *Science* 347, 2015.

64 Ryan Singel, 'Netflix Spilled Your Brokeback Mountain Secret, Lawsuit Claims', *Wired*, 17 December 2009.

65 Aliya Ram and Madhumita Murgia, 'Data Brokers: Regulators Try To Rein In The "Privacy Deathstars" ', *Financial Times*, 8 January 2019.

66 Natasha Singer, 'Data Broker Is Charged With Selling Consumers' Financial Details to "Fraudsters" ', *New York Times*, 23 December 2014.

67 Melanie Hicken, 'Data Brokers Selling Lists of Rape Victims, AIDS Patients', CNN, 19 December 2013.

68 Nitasha Tiku, 'Privacy Groups Claim Online Ads Can Target Abuse Victims', *Wired*, 27 January 2019.

69 Nicole Kobie, 'Heathrow's Facial Recognition Tech Could Make Airports More Bearable', *Wired*, 18 October 2018; Gregory Wallace, 'Instead of the Boarding Pass, Bring Your Smile to the Airport', CNN, 18 September 2018.

70 Davey Alba, 'The US Government Will Be Scanning Your Face At 20 Top Airports, Documents Show', *BuzzFeed*, 11 March 2019.

71 Kaveh Waddell, 'A NASA Engineer Was Required To Unlock His Phone At The Border', *Atlantic*, 13 February 2017.

72 Daniel Victor, 'What Are Your Rights if Border Agents Want to Search Your Phone?', *New York Times*, 14 February 2017.

73 Gemma Galdon Clavell, 'Protect Rights at Automated Borders', *Nature* 543, 2017.

74 Olivia Solon, ' "Surveillance Society": Has Technology at the US-Mexico Border Gone Too Far?', *Guardian*, 13 June 2018.

75 Douglas Heaven, 'An AI Lie Detector Will Interrogate Travellers at Some EU Borders', *New Scientist*, 31 October 2018.

76 Dylan Curran, 'Are You Ready? Here Is All The Data Facebook And Google Have On You', *Guardian*, 30 March 2018.

77 John Naughton, 'More Choice on Privacy Just Means More Chances to Do What's Best for Big Tech', *Guardian*, 8 July 2018.

78 Alex Hern, 'Privacy Policies of Tech Giants "Still Not GDPR-Compliant"', *Guardian*, 5 July 2018.

79 Logan Koepke, '"We Can Change These Terms at Anytime": The Detritus of Terms of Service Agreements', *Medium*, 18 January 2015.

80 John Naughton, 'More Choice on Privacy Just Means More Chances to Do What's Best for Big Tech'.

81 Arwa Mahdawi, 'Spotify Can Tell If You're Sad. Here's Why That Should Scare You', *Guardian*, 16 September 2018.

82 Alfred Ng, 'With Smart Sneakers, Privacy Risks Take a Great Leap', CNET, 13 February 2019.

83 Christopher Mims, 'Here Comes "Smart Dust," The Tiny Computers That Pull Power From The Air', *Wall Street Journal*, 8 November 2018.

CHAPTER TWO

1 Shoshana Zuboff, *The Age of Surveillance Capitalism* (London: Profile Books, 2019), Ch 3.

2 Samuel Gibbs and Alex Hern, 'Google at 20: How Two "Obnoxious" Students Changed the Internet', *Guardian*, 24 September 2018.

3 John Battelle, 'The Birth of Google', *Wired*, 1 August 2005.

4 Samuel Gibbs and Alex Hern, 'Google at 20: How Two "Obnoxious" Students Changed the Internet'.

5 Steven Levy, *In the Plex. How Google Thinks, Works, and Shapes Our Lives* (New York: Simon & Schuster, 2011), 77–78.

6 Google's 2004 Annual Report to the United States Securities and Exchange Commission (https://www.sec.gov/Archives/edgar/data/1288776/000119312505065298/d10k.htm)

7 Sergey Brin and Lawrence Page, 'The Anatomy of a Large-Scale Hypertextual Web Search Engine', *Computer Networks and ISDN Systems* 30, 1998.

8 Steven Levy, *In the Plex. How Google Thinks, Works, and Shapes Our Lives*, 82.

9 Samuel Gibbs and Alex Hern, 'Google at 20: How Two "Obnoxious" Students Changed the Internet'.

10 Alphabet Inc. 2019 Annual Report to the United States Securities and Exchange Commission (https://abc.xyz/investor/static/pdf/20200204_alphabet_10K.pdf?cache=cdd6dbf)

11 Richard Graham, 'Google and Advertising: Digital Capitalism in the Context of Post-Fordism, the Reification of Language, and the Rise of Fake News', *Palgrave Communications* 3, 2017, 2.

12 Jennifer Lee, 'Postcards From Planet Google', *New York Times*, 28 November 2002.

13 Jennifer Lee, 'Postcards From Planet Google'.

14 Krishna Bharat, Stephen Lawrence and Meham Sahami, 'Generating User Information for Use in Targeted Advertising' (2003).

15 Steven Levy, *In the Plex. How Google Thinks, Works, and Shapes Our Lives*, 330–336.

16 Shoshana Zuboff, *The Age of Surveillance Capitalism*, 87–92.

17 Steven Levy, *In the Plex. How Google Thinks, Works, and Shapes Our Lives*, 68.

18 Douglas Edwards, *I'm Feeling Lucky: The Confessions of Google Employee Number 59* (Houghton Mifflin Harcourt, 2011), 340.

19 Shoshana Zuboff, *The Age of Surveillance Capitalism*, 89.

20 Louise Matsakis, 'The WIRED Guide to Your Personal Data (and Who Is Using It)', *Wired*, 15 February 2019.

21 'Privacy Online: Fair Information Practices in the Electronic Marketplace. A Report to Congress' (Federal Trade Commission, 2000).

22 Shoshana Zuboff, *The Age of Surveillance Capitalism*, 112–121.

23 Bruce Schneier, *Click Here to Kill Everybody. Security and Survival in a Hyper-Connected World* (New York: W. W. Norton & Company, 2018), 65.

24 Babu Kurra, 'How 9/11 Completely Changed Surveillance in U.S.', *Wired*, 11 September 2011.

25 Edward Snowden, *Permanent Record* (Macmillan, 2019).

26 To learn more about state surveillance in the United States and the revelations of Edward Snowden, I recommend reading Barton Gellman's detailed account in *Dark Mirror*, in addition to Snowden's autobiography. Barton Gellman, *Dark Mirror* (London: Bodley Head, 2020).

27 Edward Snowden, *Permanent Record*, 223–224.

28 Edward Snowden, *Permanent Record*, 278–279.

29 Edward Snowden, *Permanent Record*, 163.

30 Edward Snowden, *Permanent Record*, 225.

31 Edward Snowden, *Permanent Record*, 167–168.

32 Michael Isikoff, 'NSA Program Stopped No Terror Attacks, Says White House Panel Member', NBC News, 19 December 2013.

33 Charlie Savage, 'Declassified Report Shows Doubts About Value of N.S.A.'s Warrantless Spying', *New York Times*, 25 April 2015.

34 Charlie Savage, *Power Wars. Inside Obama's Post-9/11 Presidency* (New York: Little, Brown and Company, 2015), 162–223.

35 'Report on the President's Surveillance Program' (2009), 637.

36 For more on why mass surveillance is not the right approach to preventing terrorism, see Bruce Schneier, *Data and Goliath*, 135–139.

37 James Glanz and Andrew W. Lehren, 'NSA Spied on Allies, Aid Groups and Businesses', *New York Times*, 21 December 2013.

38 Julia Angwin, Jeff Larson, Charlie Savage, James Risen, Henrik Moltke and Laura Poitras, 'NSA Spying Relies on AT&T's "Extreme Willingness to Help"', *ProPublica*, 15 August 2015.

39 Jedediah Purdy, 'The Anti-Democratic Worldview of Steve Bannon and Peter Thiel', *Politico*, 30 November 2016.

40 Sam Biddle, 'How Peter Thiel's Palantir Helped the NSA Spy on the Whole World', *Intercept*, 22 February 2017.

41 Bruce Schneier, *Click Here to Kill Everybody. Security and Survival in a Hyper-Connected World*, 65.

42 Sam Levin, 'Tech Firms Make Millions from Trump's Anti-Immigrant Agenda, Report Finds', *Guardian*, 23 October 2018.

43 Amanda Holpuch, 'Trump's Separation of Families Constitutes Torture, Doctors Find', *Guardian*, 25 February 2020.

44 Jason Wilson, 'Private Firms Provide Software and Information to Police, Documents Show', *Guardian*, 15 October 2020.

45 Alfred Ng, 'Google Is Giving Data to Police Based on Search Keywords, Court Docs Show', *CNET*, 8 October 2020.

46 'The Government Uses "Near Perfect Surveillance" Data on Americans', *New York Times*, 7 February 2020.

47 Toby Helm, 'Patient Data From GP Surgeries Sold to US Companies', *Observer*, 7 December 2019.

48 Juliette Kayyem, 'Never Say "Never Again"', *Foreign Policy*, 11 September 2012.

49 Bobbie Johnson, 'Privacy No Longer a Social Norm, Says Facebook Founder', *Guardian*, 10 January 2010.

50 Alyson Shontell, 'Mark Zuckerberg Just Spent More Than $30 Million Buying 4 Neighboring Houses So He Could Have Privacy', *Business Insider*, 11 October 2013.

51 Bobbie Johnson, 'Facebook Privacy Change Angers Campaigners', *Guardian*, 10 December 2009.

52 My thanks to Judith Curthoys for this example. As Ellen Judson pointed out to me, a master at Cambridge University also kept a banned dog as a 'very large cat' (https://www.bbc.co.uk/news/uk-england-cambridgeshire-28966001).

53 Harry Cockburn, 'The UK's Strangest Laws That Are Still Enforced', *Independent*, 8 September 2016.

54 Nick Statt, 'Facebook CEO Mark Zuckerberg Says the "Future is Private"', *Verge*, 30 April 2019.

55 Sam Biddle, 'In Court, Facebook Blames Users for Destroying Their Own Right to Privacy', *Intercept*, 14 June 2014.

56 Roxanne Bamford, Benedict Macon-Cooney, Hermione Dace and Chris Yiu, 'A Price Worth Paying: Tech, Privacy and the Fight Against Covid-19' (Tony Blair Institute for Global Change, 2020).

57 Barrington Moore, *Privacy. Studies in Social and Cultural History* (Armonk, New York: M. E. Sharpe, 1984).

CHAPTER THREE

1 Tim Wu, *The Attention Merchants* (Atlantic Books, 2017); James Williams, *Stand Out of Our Light. Freedom and Resistance in the Attention Economy* (Cambridge: Cambridge University Press, 2018).

2 Alex Hern, 'Netflix's Biggest Competitor? Sleep', *Guardian*, 18 April 2017.

3 Although in this book I am following the common usage of calling people who break into security systems 'hackers', the more precise term is 'crackers'. Crackers are malicious hackers. See Richard Stallman, 'On Hacking' (https://stallman.org/articles/on-hacking.html)

4 Oliver Ralph, 'Insurance and the Big Data Technology Revolution', *Financial Times*, 24 February 2017.

5 Dave Smith and Phil Chamberlain, 'On the Blacklist: How Did the UK's Top Building Firms Get Secret Information on Their Workers?', *Guardian*, 27 February 2015.

6 Rupert Jones, 'Identity Fraud Reaching Epidemic Levels, New Figures Show', *Guardian*, 23 August 2017.

7 Kaleigh Rogers, 'Let's Talk About Mark Zuckerberg's Claim That Facebook "Doesn't Sell Data"', *Motherboard*, 11 April 2018.

8 Charlie Warzel and Ash Ngu, 'Google's 4,000-Word Privacy Policy Is a Secret History of the Internet', *New York Times*, 10 July 2019.

9 Rainer Forst, 'Noumenal Power', *Journal of Political Philosophy* 23, 2015.

10 M. Weber, *Economy and Society* (Berkeley: University of California Press, 1978), 53.

11 Bertrand Russell, *Power. A New Social Analysis* (Routledge, 2004), 4.

12 Michel Foucault, *Discipline and Punish* (London: Penguin Books, 1977); Nico Stehr and Marian T. Adolf, 'Knowledge/Power/Resistance', *Society* 55, 2018.

13 Hubert Dreyfus and Paul Rabinow, *Michel Foucault. Beyond Structuralism and Hermeneutics* (University of Chicago Press, 1982), 212.

14 Steven Lukes, *Power. A Radical View* (Red Globe Press, 2005).

15 Simon Parkin, 'Has Dopamine Got Us Hooked on Tech?', *Guardian*, 4 March 2018.

16 https://www.britannica.com/topic/Stasi

17 Andrea Peterson, 'Snowden Filmmaker Laura Poitras: "Facebook is a Gift to Intelligence Agencies"', *Washington Post*, 23 October 2014.

18 Robert Booth, Sandra Laville and Shiv Malik, 'Royal Wedding: Police Criticised for Pre-Emptive Strikes Against Protestors', *Guardian*, 29 April 2011.

19 Tae Kim, 'Warren Buffett Believes This Is "the Most Important Thing" to Find in a Business', CNBC, 7 May 2018.

20 Associated Press, 'Google Records Your Location Even When You Tell It Not To', *Guardian*, 13 August 2018.

21 Frank Tang, 'China Names 169 People Banned From Taking Flights or Trains Under Social Credit System', *South China Morning Post*, 2 June 2018.

22 Simina Mistreanu, 'Life Inside China's Social Credit Laboratory', *Foreign Policy*, 3 April 2018.

23 Orange Wang, 'China's Social Credit System Will Not Lead to Citizens Losing Access to Public Services, Beijing Says', *South China Morning Post*, 19 July 2019.

24 Nectar Gan, 'China Is Installing Surveillance Cameras Outside People's Front Doors . . . and Sometimes Inside Their Homes', *CNN Business*, 28 April 2020.

25 Hill's article mentions a list of other companies that score consumers and how to contact them to ask for your data. Kashmir Hill, 'I Got Access to My Secret Consumer Score. Now You Can Get Yours, Too', *New York Times*, 4 November 2019.

26 Edward Snowden, *Permanent Record*, 196–197.

27 Jamie Susskind, *Future Politics. Living Together in a World Transformed by Tech* (Oxford University Press, 2018), 103–107.

28 Jamie Susskind, *Future Politics. Living Together in a World Transformed by Tech*, 172.

29 I got the idea that manipulation makes the victim complicit in her own victimization from philosopher Robert Noggle. Workshop on Behavioural Prediction and Influence, 'The Moral Status of "Other Behavioral Influences"', University of Oxford (27 September 2019).

30 Richard Esguerra, 'Google CEO Eric Schmidt Dismisses the Importance of Privacy', *Electronic Frontier Foundation*, 10 December 2009.

31 Steven Levy, *In the Plex. How Google Thinks, Works, and Shapes Our Lives*, 175.

32 There is an argument to be made that we should be allowed to hide minor transgressions (e.g. see Cressida Gaukroger, 'Privacy and the Importance of "Getting Away With It"', *Journal of Moral Philosophy* 17, 2020), but that is not one of the most important functions of privacy.

33 Carissa Véliz, 'Inteligencia artificial: ¿progreso o retroceso?', *El País*, 14 June 2019.

34 Shoshana Zuboff, *The Age of Surveillance Capitalism*, 221–225.

35 Bent Flyvbjerg, *Rationality and Power. Democracy in Practice* (Chicago University Press, 1998), 36.

36 Safiya Noble, *Algorithms of Oppression. How Search Engines Reinforce Racism* (NYU Press, 2018); Caroline Criado Perez, *Invisible Women. Exposing Data Bias in a World Designed for Men* (Vintage, 2019).

37 James Zou and Londa Schiebinger, 'AI Can Be Sexist and Racist – It's Time to Make It Fair', *Nature* 559, 2018.

38 Danny Yadron, 'Silicon Valley Tech Firms Exacerbating Income Inequality, World Bank Warns', *Guardian*, 15 January 2016.

39 https://www.energy.gov/articles/history-electric-car

40 Nick Bilton, 'Why Google Glass Broke', *New York Times*, 4 February 2015.

41 Nick Bilton, 'Why Google Glass Broke'.

42 https://about.fb.com/news/2020/09/announcing-project-aria-a-research-project-on-the-future-of-wearable-ar/

43 Steven Poole, 'Drones the Size of Bees – Good or Evil?', *Guardian*, 14 June 2013.

44 Rose Eveleth, 'The Biggest Lie Tech People Tell Themselves – and the Rest of Us', *Vox*, 8 October 2019.

45 James Williams, *Stand Out of Our Light. Freedom and Resistance in the Attention Economy.*

46 Gmail no longer scans our emails for the purposes of personalized ads, but it did so up until 2017, and third-party apps continue to do so (though you can remove that access through your settings). Christopher Wylie, *Mindf*ck. Inside Cambridge Analytica's Plot to Break the World* (Profile Books, 2019), 15. Alex Hern, 'Google Will Stop Scanning Content of Personal Emails', *Guardian*, 26 June 2017. Kaya Yurieff, 'Google Still Lets Third-Party Apps Scan Your Gmail Data', *CNN Business*, 20 September 2018.

47 Christopher Wylie, *Mindf*ck. Inside Cambridge Analytica's Plot to Break the World*, 15.

48 Christopher Wylie, *Mindf*ck. Inside Cambridge Analytica's Plot to Break the World*, 16.

49 George Orwell, *Politics and the English Language* (Penguin, 2013).

50 'Nature's Language Is Being Hijacked By Technology', BBC News, 1 August 2019.

51 Christopher Wylie, *Mindf*ck. Inside Cambridge Analytica's Plot to Break the World*, 101–102.

52 Facebook allowed thousands of other developers to download the data of unknowing friends of people who had consented to use an app. They include the makers of games such as FarmVille, Tinder, and Barack Obama's presidential campaign. Facebook changed this policy in 2015. Elizabeth Dwoskin and Tony Romm, 'Facebook's Rules for Accessing User Data Lured More Than Just Cambridge Analytica', *Washington Post*, 20 March 2018.

53 Christopher Wylie, *Mindf*ck. Inside Cambridge Analytica's Plot to Break the World*, 110–111.

54 Brittany Kaiser, *Targeted. My Inside Story of Cambridge Analytica and How Trump, Brexit and Facebook Broke Democracy*, Ch 9, Ch 13.

55 Christopher Wylie, *Mindf*ck. Inside Cambridge Analytica's Plot to Break the World*, Ch 7.

56 https://www.channel4.com/news/cambridge-analytica-revealed-trumps-election-consultants-filmed-saying-they-use-bribes-and-sex-workers-to-entrap-politicians-investigation

57 Alexander Nix (the CEO) and Mark Turnbull offered some of these services to someone they thought would be their client in the Channel 4 undercover investigation. Cambridge Analytica accused Channel 4 of entrapment leading Nix to amend his statements regarding their activities. Emma Graham-Harrison, Carole Cadwalladr and Hilary Osborne, 'Cambridge Analytica Boasts of Dirty Tricks to Swing Elections', *Guardian*, 19 March 2018: https://www.theguardian.com/uk-news/2018/mar/19/cambridge-analytica-execs-boast-dirty-tricks-honey-traps-elections. More recently, the UK Insolvency Service has banned Nix from serving as a company director for seven years on account of offering the

'potentially unethical' services mentioned before to prospective clients. Rob Davies, 'Former Cambridge Analytica Chief Receives Seven-Year Directorship Ban', *Guardian*, 24 September 2020.

58 Christopher Wylie, *Mindf*ck. Inside Cambridge Analytica's Plot to Break the World*, Ch 8. The Information Commissioner's Office in the UK has recently released a report of its investigation into Cambridge Analytica. Some commentators interpreted the report to be exonerating the data firm from ties to Russia. In fact, the ICO simply stated that matters related to 'possible Russia-located activity' fell beyond its remit. Other commentators on the report gave the impression that the ICO had found that Cambridge Analytica had not been actively involved in the Brexit referendum. In fact, the report mentions the activity of AggregateIQ, a Canada-based company that was linked to Cambridge Analytica and its parent company, SCL, and that was involved with the Leave campaign.

59 Christopher Wylie, *Mindf*ck. Inside Cambridge Analytica's Plot to Break the World*, 244.

60 ICO report (ICO/O/ED/L/RTL/0181), p.16: 'We understand from witness evidence that AIQ played a significant role in the deployment of targeted advertisement, leveraging their expertise in this digital marketing in order to assist SCL. There was a range of evidence that demonstrated a very close relationship between AIQ and SCL (such as evidence that described AIQ as the Canadian branch of SCL and evidence that Facebook invoices to AIQ for advertising were paid directly by SCL). However, AIQ has consistently denied having a closer relationship beyond that between a software developer and their client. Mr Silvester (a director/owner of AIQ) has stated that in 2014 SCL "asked us to create SCL Canada but we declined".'

61 Amber Macintyre, 'Who's Working for Your Vote?', *Tactical Tech*, 29 November 2018.

62 Lorenzo Franceschi-Bicchierai, 'Russian Facebook Trolls Got Two Groups of People to Protest Each Other in Texas', *Motherboard*, 1 November 2017.

63 Gary Watson, 'Moral Agency', *The International Encyclopedia of Ethics* (2013); John Christman, 'Autonomy in Moral and Political Philosophy', in Edward N. Zalta (ed.), *The Stanford Encyclopedia of Philosophy* (2015).

64 Myrna Oliver, 'Legends Nureyev, Gillespie Die: Defector Was One of Century's Great Dancers', *Los Angeles Times*, 7 January 1993.

65 Jonathon W. Penney, 'Chilling Effects: Online Surveillance and Wikipedia Use', *Berkeley Technology Law Journal* 31, 2016.

66 Karina Vold and Jess Whittlestone, 'Privacy, Autonomy, and Personalised Targeting: Rethinking How Personal Data Is Used', in Carissa Véliz (ed.), *Data, Privacy, and the Individual* (Center for the Governance of Change, IE University, 2019).

67 Hamza Shaban, 'Google for the First Time Outspent Every Other Company to Influence Washington in 2017', *Washington Post*, 23 January 2018.

68 Caroline Daniel and Maija Palmer, 'Google's Goal: To Organise Your Daily Life', *Financial Times*, 22 May 2007.

69 Holman W. Jenkins, 'Google and the Search for the Future', *Wall Street Journal*, 14 August 2010.

70 Carissa Véliz, 'Privacy is a Collective Concern', *New Statesman*, 22 October 2019.

71 Carissa Véliz, 'Data, Privacy & the Individual' (Madrid: Center for the Governance of Change, IE University, 2020).

72 Kristen V. Brown, 'What DNA Testing Companies' Terrifying Privacy Policies Actually Mean'.

73 Jody Allard, 'How Gene Testing Forced Me to Reveal My Private Health Information', *Vice*, 27 May 2016.

74 https://blog.23andme.com/health-traits/sneezing-on-summer-solstice/

75 S. L. Schilit and A. Schilit Nitenson, 'My Identical Twin Sequenced our Genome', *Journal of Genetic Counseling* 26, 2017.

76 Lydia Ramsey and Samantha Lee, 'Our DNA is 99.9% the Same as the Person Next to Us – and We're Surprisingly Similar to a Lot of Other Living Things', *Business Insider*, 3 April 2018.

77 Jocelyn Kaiser, 'We Will Find You: DNA Search Used to Nab Golden State Killer Can Home In On About 60% of White Americans', *Science Magazine*, 11 October 2018.

78 Tamara Khandaker, 'Canada Is Using Ancestry DNA Websites To Help It Deport People', *Vice*, 26 July 2018.

79 Jocelyn Kaiser, 'We Will Find You: DNA Search Used to Nab Golden State Killer Can Home In On About 60% of White Americans'.

80 Matthew Shaer, 'The False Promise of DNA Testing', *Atlantic*, June 2016.

81 Brendan I. Koerner, 'Your Relative's DNA Could Turn You Into a Suspect', *Wired*, 13 October 2015.

82 https://www.innocenceproject.org/overturning-wrongful-convictions-involving-flawed-forensics/

83 Keith Hampton, Lee Rainie, Weixu Lu, Maria Dwyer, Inyoung Shin and Kristen Purcell, 'Social Media and the "Spiral of Silence" ', Pew Research Center, (2014); Elizabeth Stoycheff, 'Under Surveillance: Examining Facebook's Spiral of Silence Effects in the Wake of NSA Monitoring', *Journalism & Mass Communication Quarterly* 93, 2016.

84 Javier de la Cueva, personal communication.

85 Kieron O'Hara and Nigel Shadbolt, 'Privacy on the Data Web', *Communications of the ACM* 53, 2010.

86 Robert B. Talisse, 'Democracy: What's It Good For?', *Philosophers' Magazine* 89, 2020.

87 'A Manifesto for Renewing Liberalism', *The Economist*, 15 September 2018.

88 Michael J. Abramowitz, 'Democracy in Crisis', *Freedom in the World* (2018).

89 *The Economist* Intelligence Unit, 'Democracy Index 2019. A Year of Democratic Setbacks and Popular Protest' (2019).

90 *Democracy Under Lockdown*, Freedom House, October 2020. https://freedomhouse.org/article/new-report-democracy-under-lockdown-impact-covid-19-global-freedom

91 https://api.parliament.uk/historic-hansard/commons/1947/nov/11/parliament-bill

92 John Stuart Mill, *On Liberty* (Indianapolis: Hackett Publishing Company, 1978), Ch 3.

93 Thanks to Mauricio Suárez for reminding me of democratic peace theory, and to Antonio Diéguez for reminding me of Karl Popper's argument.

94 Karl Popper, *The Open Society and Its Enemies* (Routledge, 2002), 368.

95 George Orwell, *Fascism and Democracy* (Penguin, 2020), 6.

96 Steven Levitsky and Daniel Ziblatt, *How Democracies Die* (Penguin, 2018), 3.

97 Jonathan Wolff, 'The Lure of Fascism', *Aeon*, 14 April 2020.

98 Hidalgo argues that we should get rid of political representatives and instead have our digital assistants vote on our behalf. He claims this is a type of 'direct democracy', but I find that questionable – it could be argued that we would merely be exchanging our human representatives for digital ones. (Not that I think direct democracy is better than representative democracy.) https://www.ted.com/talks/cesar_hidalgo_a_bold_idea_to_replace_politicians

99 Sam Wolfson, 'For My Next Trick: Dynamo's Mission to Bring Back Magic', *Guardian*, 26 April 2020.

100 Cecilia Kang and Kenneth P. Vogel, 'Tech Giants Amass a Lobbying Army for an Epic Washington Battle', *New York Times*, 5 June 2019; Tony Romm, 'Tech Giants Led by Amazon, Facebook and Google Spent Nearly Half a Billion on Lobbying Over the Last Decade', *Washington Post*, 22 January 2020.

101 Rana Foroohar, 'Year in a Word: Techlash', *Financial Times*, 16 December 2018.

CHAPTER FOUR

1 Tom Douglas, 'Why the Health Threat From Asbestos Is Not a Thing of the Past', *The Conversation*, 21 December 2015.

2 Bruce Schneier, 'Data is a Toxic Asset, So Why Not Throw it Out?', CNN, 1 March 2016.

3 Tom Lamont, 'Life After the Ashley Madison Affair', *Observer*, 27 February 2016.

4 Rob Price, 'An Ashley Madison User Received a Terrifying Blackmail Letter', *Business Insider*, 22 January 2016.

5 Chris Baraniuk, 'Ashley Madison: "Suicides" Over Website Hack', BBC News, 24 August 2015; 'Pastor Outed on Ashley Madison Commits Suicide', Laurie Segall, CNN, 8 September 2015.

6 José Antonio Hernández, 'Me han robado la identidad y estoy a base de lexatín; yo no soy una delincuente', *El País*, 24 August 2016.

7 Siân Brooke and Carissa Véliz, 'Views on Privacy. A Survey', *Data, Privacy & the Individual* (Center for the Governance of Change, IE University, 2020).

8 Alex Hern, 'Hackers Publish Private Photos From Cosmetic Surgery Clinic'.

9 Zoe Kleinman, 'Therapy Patients Blackmailed for Cash After Clinic Data Breach', BBC News, 26 October 2020.

10 Isabel Valdés, 'La Fiscalía investiga el suicidio de una empleada de Iveco tras la difusión de un vídeo sexual', *El País*, 30 May 2019.

11 'How WhatsApp Helped Turn an Indian Village Into a Lynch Mob', BBC News, 18 July 2018.

12 'Stalker "Found Japanese Singer Through Reflection in Her Eyes"', BBC News, 10 October 2019.

13 Kashmir Hill, 'Wrongfully Accused by an Algorithm', *New York Times*, 24 June 2020.

14 Oren Liebermann, 'How a Hacked Phone May Have Led Killers to Khashoggi', *CNN*, 20 January 2019.

15 Siân Brooke and Carissa Véliz, 'Views on Privacy. A Survey'.

16 Olivia Solon, 'Ashamed to Work in Silicon Valley: How Techies Became the New Bankers', *Guardian*, 8 November 2017.

17 'FTC Imposes $5 Billion Penalty and Sweeping New Privacy Restrictions on Facebook', FTC Press Release, 24 July 2019.

18 'Facebook Fined £500,000 for Cambridge Analytica Scandal', BBC News, 25 October 2018.

19 'British Airways Faces Record £183m Fine for Data Breach', BBC News, 8 July 2019.

20 David E. Sanger, 'Hackers Took Fingerprints of 5.6 Million U.S. Workers, Government Says', *New York Times*, 23 September 2015.

21 Edward Wong, 'How China Uses LinkedIn to Recruit Spies Abroad', *New York Times*, 27 August 2019.

22 Jordi Pérez Colomé, 'Por qué China roba datos privados de decenas de millones de estadounidenses', *El País*, 17 February 2020.

23 On its dedicated settlement website, the company notes that 'Equifax denied any wrongdoing, and no judgment or finding of wrongdoing has been made.' https://www.equifaxbreachsettlement.com. Equifax agreed to pay $700m as part of a settlement with the Federal Trade Commission (FTC). 'Equifax failed to take basic steps that may have prevented the breach', said the FTC's chairman Joe Simons. 'Equifax to Pay Up to $700m to Settle Data Breach', BBC News, 22 July 2019. The class action lawsuit against Equifax can be found at http://securities.stanford.edu/filings-documents/1063/EI00_15/2019128_r01x_17CV03463.pdf

24 Charlie Warzel, 'Chinese Hacking Is Alarming. So Are Data Brokers', *New York Times*, 10 February 2020.

25 Stuart A. Thompson and Charlie Warzel, 'Twelve Million Phones, One Dataset, Zero Privacy', *New York Times*, 19 December 2019.

26 Stuart A. Thompson and Charlie Warzel, 'How to Track President Trump', *New York Times*, 20 December 2019.

27 At the time of writing, Oracle and Walmart are jointly bidding for TikTok. Given that Oracle owns and works with more than 80 data brokers, it doesn't sound like

good news for privacy. Oracle claims to sell data on more than 300m people globally, with 30,000 data points per individual (including shopping behaviour, financial transactions, social media behaviour, demographic information and more), covering '"over 80 per cent of the entire US internet population"'. Aliya Ram and Madhumita Murgia, 'Data Brokers: Regulators Try To Rein In The "Privacy Deathstars"'.

28 Devin Coldewey, 'Grindr Sends HIV Status to Third Parties, and Some Personal Data Unencrypted', *TechCrunch*, 2 April 2018. The Norwegian Consumer Council Investigation into Grindr's consent mechanism resulted in Grindr rejecting a number of the report's suggestions while welcoming a best practice discussion: 'Grindr and Twitter Face "Out of Control" Complaint', BBC News, 14 January 2020.

29 Echo Wang and Carl O'Donnell, 'Behind Grindr's Doomed Hookup in China, a Data Misstep and Scramble to Make Up', Reuters, 22 May 2019.

30 Casey Newton, 'How Grindr Became a National Security Issue', *Verge*, 28 March 2019.

31 Jeremy Hsu, 'The Strava Heat Map and the End of Secrets', *Wired*, 29 January 2018.

32 Colin Lecher, 'Strava Fitness App Quietly Added a New Opt-Out for Controversial Heat Map', *Verge*, 1 March 2018.

33 Pablo Guimón, ' "Brexit Wouldn't Have Happened Without Cambridge Analytica" ', *El País*, 27 March 2018.

34 Alex Hern, 'Facebook "Dark Ads" Can Swing Political Opinions, Research Shows', *Guardian*, 31 July 2017; Timothy Revell, 'How to Turn Facebook Into a Weaponised AI Propaganda Machine', *New Scientist*, 28 July 2017; Sue Halpern, 'Cambridge Analytica and the Perils of Psychographics', *New Yorker*, 30 March 2018.

35 Angela Chen and Alessandra Potenza, 'Cambridge Analytica's Facebook Data Abuse Shouldn't Get Credit for Trump', *Verge*, 20 March 2018; Kris-Stella Trump, 'Four and a Half Reasons Not to Worry That Cambridge Analytica Skewed the 2016 Election', *Washington Post*, 23 March 2018.

36 Kyle Endres, 'Targeted Issue Messages and Voting Behavior', *American Politics Research* 48, 2020.

37 The paper describes carrying out a randomized control trial 'with all users of at least 18 years of age in the United States who accessed the Facebook website on 2 November 2010'. Presumably Facebook assumed that their study was covered under its terms and conditions – an extremely questionable assumption. A similar controversy ensued in 2014, when Facebook published a study on emotional contagion. Kashmir Hill, a reporter, pointed out that Facebook had added to its User Agreement the possibility of data being used for research four months after the study took place. Even then, it is arguably not the case that agreeing to terms and conditions that most people don't read can count as informed consent. Kashmir Hill, 'Facebook Added "Research" To User Agreement 4 Months After Emotion Manipulation Study', *Forbes*, 30 June 2014.

38 M. Bond, C. J. Fariss, J. J. Jones, A. D. Kramer, C. Marlow, J. E. Settle and J. H. Fowler, 'A 61-Million-Person Experiment in Social Influence and Political Mobilization', *Nature* 489, 2012.

39 Jay Caruso, 'The Latest Battleground Poll Tells Us Democrats Are Over-Correcting for 2020 – and They Can't Beat Trump That Way', *Independent*, 5 November 2019.

40 'Revealed: Trump Campaign Strategy to Deter Millions of Black Americans from Voting in 2016', Channel 4, https://www.youtube.com/watch?v=KIf5ELaOjOk

41 Hannes Grassegger, 'Facebook Says Its "Voter Button" Is Good for Turnout. But Should the Tech Giant Be Nudging Us at All?', *Observer*, 15 April 2018.

42 John Gramlich, '10 Facts About Americans and Facebook', Pew Research Center, 16 May 2019.

43 *Wired* counted twenty-one scandals just for 2018. Issie Lapowsky, 'The 21 (and Counting) Biggest Facebook Scandals of 2018', *Wired*, 20 December 2018.

44 In 2019, Zuckerberg announced Facebook would not moderate or fact-check political ads. As events unfolded, and Facebook received pressure to change its policies, Zuckerberg somewhat relented. Facebook first banned ads meant to suppress votes. More recently, the company has prohibited political ads that seek 'to delegitimize an election', including claims of voting fraud. The platform also decided to prohibit all political ads after the polls close on 3 November for an undetermined length of time. While these policies are positive, we have no guarantee, first, that they will be implemented in a reliable way, and second, that they are enough to guarantee fair play on Facebook. Cecilia Kang and Mike Isaac, 'Defiant Zuckerberg Says Facebook Won't Police Political Speech', *New York Times*, 17 October 2019. As this book was about to go to press, Facebook announced some changes shortly after more than 100 brands pulled advertising from the platform amid backlash over Facebook's policies on hate speech. Facebook said it will remove posts that incite violence or attempt to suppress voting, and affix labels on newsworthy posts that violate other policies, much like Twitter does. Kari Paul, 'Facebook Policy Changes Fail to Quell Advertiser Revolt as Coca-Cola Pulls Ads', *Guardian*, 27 June 2020.

45 Tim Wu, 'Facebook Isn't Just Allowing Lies, It's Prioritizing Them', *New York Times*, 4 November 2019.

46 Andrew Marantz, 'Why Facebook Can't Fix Itself', *New Yorker*, 12 October 2020.

47 Karen Kornbluh, Adrienne Goldstein and Eli Weiner, 'New Study by Digital New Deal Finds Engagement With Deceptive Outlets Higher on Facebook Today Than Run-Up to 2016 Election', German Marshall Fund of the United States, 12 October 2020.

48 David Smith, 'How Key Republicans Inside Facebook Are Shifting Its Politics to the Right', *Guardian*, 3 November 2019.

49 Jonathan Zittrain, 'Facebook Could Decide an Election Without Anyone Ever Finding Out', *New Statesman*, 3 June 2014.

50 Whistleblowers Chris Wylie and Brittany Kaiser both claim that Cambridge Analytica engaged in voter suppression. Donie O'Sullivan and Drew Griffin,

'Cambridge Analytica Ran Voter Suppression Campaigns, Whistleblower Claims', CNN, 17 May 2018; Brittany Kaiser, *Targeted. My Inside Story of Cambridge Analytica and How Trump, Brexit and Facebook Broke Democracy*, 231.

51 In response to calls for more transparency, Facebook opened an online library of all the advertisements on its platform. Journalists and researchers, however, complain that the tool is 'so plagued by bugs and technical constraints that it is effectively useless as a way to comprehensively track political advertising'. Matthew Rosenberg, 'Ad Tool Facebook Built to Fight Disinformation Doesn't Work as Advertised', *New York Times*, 25 July 2019.

52 John Stuart Mill, *Collected Works of John Stuart Mill* (University of Toronto Press, 1963), vol. 21, 262.

53 Thomas Nagel, 'Concealment and Exposure', *Philosophy and Public Affairs* 27, 1998.

54 Anna Lauren Hoffman, 'Facebook is Worried About Users Sharing Less – But it Only Has Itself to Blame', *Guardian*, 19 April 2016.

55 Thomas Nagel, 'Concealment and Exposure'.

56 Edwin Black, *IBM and the Holocaust* (Washington, DC: Dialog Press, 2012), Ch 11.

57 William Seltzer and Margo Anderson, 'The Dark Side of Numbers: The Role of Population Data Systems in Human Rights Abuses', *Social Research* 68, 2001.

58 William Seltzer and Margo Anderson, 'The Dark Side of Numbers: The Role of Population Data Systems in Human Rights Abuses'.

59 Hans de Zwart, 'During World War II, We Did Have Something to Hide', *Medium*, 30 April 2015.

60 Thomas Douglas and Lauren Van den Borre, 'Asbestos Neglect: Why Asbestos Exposure Deserves Greater Policy Attention', *Health Policy* 123, 2019.

CHAPTER FIVE

1 Fiona Harvey, 'Ozone Layer Finally Healing After Damage Caused by Aerosols, UN Says', *Guardian*, 5 November 2018.

2 'Update Report Into Adtech and Real Time Bidding' (United Kingdom: Information Commissioner's Office, 2019).

3 Jesse Frederik and Maurits Martijn, 'The New Dot Com Bubble Is Here: It's Called Online Advertising ', *Correspondent*, 6 November 2019.

4 Keach Hagey, 'Behavioral Ad Targeting Not Paying Off for Publishers, Study Suggests', *Wall Street Journal*, 29 May 2019.

5 Laura Bassett, 'Digital Media Is Suffocating – and It's Facebook and Google's Fault', *American Prospect*, 6 May 2019.

6 Natasha Lomas, 'The Case Against Behavioral Advertising Is Stacking Up', *TechCrunch*, 20 January 2019.

7 Mark Weiss, 'Digiday Research: Most Publishers Don't Benefit From Behavioral Ad Targeting', *Digiday*, 5 June 2019.

8 Jessica Davies, 'After GDPR, The New York Times Cut Off Ad Exchanges in Europe – and Kept Growing Ad Revenue', *Digiday*, 16 January 2019.

9 Tiffany Hsu, 'The Advertising Industry Has a Problem: People Hate Ads', *New York Times*, 28 October 2019.

10 David Ogilvy, *Confessions of an Advertising Man* (Harpenden: Southbank Publishing, 2013), 17, 114.

11 Louise Matsakis, 'Online Ad Targeting Does Work – As Long As It's Not Creepy', *Wired*, 11 May 2018; Tami Kim, Kate Barasz and Leslie K. John, 'Why Am I Seeing This Ad? The Effect of Ad Transparency on Ad Effectiveness', *Journal of Consumer Research* 45, 2019.

12 Rani Molla, 'These Publications Have the Most to Lose From Facebook's New Algorithm Changes', *Vox*, 25 January 2018.

13 Emily Bell, 'Why Facebook's News Feed Changes Are Bad News For Democracy', *Guardian*, 21 January 2018; Dom Phillips, 'Brazil's Biggest Newspaper Pulls Content From Facebook After Algorithm Change', *Guardian*, 8 February 2018.

14 Gabriel Weinberg, 'What If We All Just Sold Non-Creepy Advertising?', *New York Times*, 19 June 2019.

15 David Ogilvy, *Confessions of an Advertising Man*, 168, 112, 127.

16 Chloé Michel, Michelle Sovinsky, Eugenio Proto and Andrew Oswald, 'Advertising as a Major Source of Human Dissatisfaction: Cross-National Evidence on One Million Europeans', in M. Rojas (ed.), *The Economics of Happiness* (Springer, 2019).

17 'Economic Impact of Advertising in the United States' (IHS Economics and Country Risk, 2015).

18 'United States of America – Contribution of Travel and Tourism to GDP as a Share of GDP' (Knoema, 2018).

19 'Something Doesn't Ad Up About America's Advertising Market', *The Economist*, 18 January 2018.

20 Eli Rosenberg, 'Quote: The Ad Generation', *The Atlantic*, 15 April 2011.

21 'Something Doesn't Ad Up About America's Advertising Market'.

22 Robert O'Harrow Jr, 'Online Firm Gave Victim's Data to Killer', *Chicago Tribune*, 6 January 2006.

23 Natasha Singer, 'Data Broker Is Charged With Selling Consumers' Financial Details to "Fraudsters"'.

24 David A. Hoffman, 'Intel Executive: Rein In Data Brokers', *New York Times*, 15 July 2019.

25 Elizabeth Dwoskin, 'FTC: Data Brokers Can Buy Your Bank Account Number for 50 Cents', *Wall Street Journal*, 24 December 2014; Julia Angwin, *Dragnet Nation* (New York: Times Books, 2014), 7.

26 Joana Moll, 'The Dating Brokers: An Autopsy of Online Love', October 2018.

27 Alex Hern, 'Apple Contractors "Regularly Hear Confidential Details" on Siri Recordings', *Guardian*, 26 July 2019; Alex Hern, 'Facebook Admits Contractors Listened to Users' Recordings Without Their Knowledge', *Guardian*, 14 August 2019; Joseph Cox, 'Revealed: Microsoft Contractors Are Listening to Some Skype

Calls', *Motherboard*, 7 August 2019; Austin Carr, Matt Day, Sarah Frier and Mark Gurman, 'Silicon Valley Is Listening to Your Most Intimate Moments', *Bloomberg Businessweek*, 11 December 2019; Alex Hern, 'Apple Whistleblower Goes Public Over "Lack of Action"', *Guardian*, 20 May 2020.

28   Nigel Shadbolt and Roger Hampson, *The Digital Ape. How to Live (in Peace) with Smart Machines* (Oxford University Press, 2019), 318.

29   Gabriel J. X. Dance, Michael LaForgia and Nicholas Confessore, 'As Facebook Raised a Privacy Wall, It Carved an Opening for Tech Giants'.

30   Shoshana Zuboff, *The Age of Surveillance Capitalism*, 138–155.

31   I take this example from an interview with Aaron Roth (he used the Trump campaign to illustrate the method): https://twimlai.com/twiml-talk-132-differential-privacy-theory-practice-with-aaron-roth/

32   Rachel Metz, 'The Smartphone App That Can Tell You're Depressed Before You Know it Yourself', *MIT Technology Review*, 15 October 2018.

33   Michal Kosinski, David Stillwell and Thore Graepel, 'Private Traits and Attributes Are Predictable From Digital Records of Human Behavior', *PNAS* 110, 2013.

34   Christopher Burr and Nello Cristianini, 'Can Machines Read our Minds?', *Minds and Machines* 29, 2019.

35   Michal Kosinski, David Stillwell and Thore Graepel, 'Private Traits and Attributes Are Predictable From Digital Records of Human Behavior'.

36   Alexis Kramer, 'Forced Phone Fingerprint Swipes Raise Fifth Amendment Questions', *Bloomberg Law*, 7 October 2019.

37   Jack M. Balkin, 'Information Fiduciaries and the First Amendment', *UC Davis Law Review* 49, 2016; Jonathan Zittrain, 'How to Exercise the Power You Didn't Ask For', *Harvard Business Review*, 19 September 2018.

38   Alice MacLachlan, 'Fiduciary Duties and the Ethics of Public Apology', *Journal of Applied Philosophy* 35, 2018.

39   Lina Khan and David E. Pozen, 'A Skeptical View of Information Fiduciaries', *Harvard Law Review* 133, 2019.

40   Bruce Schneier, *Click Here to Kill Everybody. Security and Survival in a Hyper-Connected World*, 134.

41   Andy Greenberg, 'How Hacked Water Heaters Could Trigger Mass Blackouts', *Wired*, 13 August 2018. Russia caused a blackout in Ukraine in 2016 through a cyberattack. Andy Greenberg, 'New Clues Show How Russia's Grid Hackers Aimed for Physical Destruction', *Wired*, 12 September 2019.

42   Sean Lyngaas, 'Hacking Nuclear Systems Is the Ultimate Cyber Threat. Are We Prepared?', *Verge*, 23 January 2018.

43   Will Dunn, 'Can Nuclear Weapons Be Hacked?', *New Statesman*, 7 May 2018. The United States and Israel obstructed Iran's nuclear programme through a cyberattack (Stuxnet). Ellen Nakashima and Joby Warrick, 'Stuxnet Was Work of US and Israeli Experts, Officials Say', *Washington Post*, 2 June 2012. A more worrying attack would be one that tries to activate a nuclear weapon.

44 Matthew Wall, '5G: "A Cyber-Attack Could Stop the Country"', BBC News, 25 October 2018.

45 Jillian Ambrose, 'Lights Stay On Despite Cyber-Attack on UK's Electricity System', *Guardian*, 14 May 2020.

46 'WHO Reports Fivefold Increase in Cyber Attacks, Urges Vigilance' (https://www.who.int/news-room/detail/23-04-2020-who-reports-fivefold-increase-in-cyber-attacks-urges-vigilance).

47 Bruce Schneier, *Click Here to Kill Everybody. Security and Survival in a Hyper-Connected World*, 118–119.

48 Bruce Schneier, *Click Here to Kill Everybody. Security and Survival in a Hyper-Connected World*, 32–33, 168.

49 Gary Marcus, 'Total Recall: The Woman Who Can't Forget', *Wired*, 23 March 2009.

50 Viktor Mayer-Schönberger, *Delete. The Virtue of Forgetting in the Digital Age* (Princeton University Press, 2009), 39–45.

51 Viktor Mayer-Schönberger, *Delete. The Virtue of Forgetting in the Digital Age*, Ch 4.

52 I take this example from Carl Bergstrom and Jevin West's analysis of a paper that claims that an algorithm can determine whether someone is a criminal from analysing a facial image. 'Criminal Machine Learning': https://callingbullshit.org/case_studies/case_study_criminal_machine_learning.html

53 Julia Powles and Enrique Chaparro, 'How Google Determined Our Right to be Forgotten', *Guardian*, 18 February 2015.

54 Jack Nicas, 'The Police Can Probably Break Into Your Phone', *New York Times*, 21 October 2020.

55 Some of these and other good suggestions can be found in Bruce Schneier, *Data and Goliath*, Ch 13.

56 David Cole, '"We Kill People Based on Metadata"', *New York Review of Books*, 10 May 2014.

57 Evan Selinger and Woodrow Hartzog, 'What Happens When Employers Can Read Your Facial Expressions?', *New York Times*, 17 October 2019; Woodrow Hartzog and Evan Selinger, 'Facial Recognition Is the Perfect Tool for Oppression', *Medium*, 2 August 2018. David Hambling, 'The Pentagon Has a Laser That Can Identify People from a Distance – by Their Heartbeat', *MIT Technology Review*, 27 June 2019.

58 Tom Miles, 'UN Surveillance Expert Urges Global Moratorium on Sale of Spyware', Reuters, 18 June 2019.

59 Nick Hopkins and Stephanie Kirchgaessner, 'WhatsApp Sues Israeli Firm, Accusing It of Hacking Activists' Phones', *Guardian*, 29 October 2019.

60 Sarah Parcak, 'Are We Ready for Satellites That See Our Every Move?', *New York Times*, 15 October 2019.

61 Amy Maxmen, 'Surveillance Science', *Nature* 569, 2019.

62 The Mission to Create a Searchable Database of Earth's Surface (https://www.ted.com/talks/will_marshall_the_mission_to_create_a_searchable_database_of_earth_s_surface).

63 James Vincent, 'iRobot's Latest Roomba Remembers Your Home's Layout and Empties Itself', *Verge*, 6 September 2018.

64 Evan Ackerman, 'Why You Sould Be Very Skeptical of Ring's Indoor Security Drone', *IEEE Spectrum*, 25 September 2020.

65 https://about.fb.com/news/2020/09/announcing-project-aria-a-research-project-on-the-future-of-wearable-ar/

66 Adam Satariano, 'Europe's Privacy Law Hasn't Shown Its Teeth, Frustrating Advocates', *New York Times*, 27 April 2020.

67 Steve Lohr, 'Forget Antitrust Laws. To Limit Tech, Some Say a New Regulator Is Needed', New York Times, 22 October 2020.

68 I was prompted to think more about the importance of diplomacy by an online conversation I had with Tom Fletcher, Zeid Ra'ad, and Mike Wooldridge, and by Tom Fletcher's book, *The Naked Diplomat*. To see the conversation: https://www.youtube.com/watch?v=LnV8iUoCLpg

69 Carissa Véliz, 'You've Heard of Tax Havens. After Brexit, the UK Could Become a "Data Haven"', *Guardian*, 17 October 2020.

70 Lois Beckett, 'Under Digital Surveillance: How American Schools Spy on Millions of Kids', *Guardian*, 22 October 2019.

71 Tristan Louis, 'How Much Is a User Worth?', *Forbes*, 31 August 2013.

72 James H. Wilson, Paul R. Daugherty and Chase Davenport, 'The Future of AI Will Be About Less Data, Not More', *Harvard Business Review*, 14 January 2019.

73 Bruce Schneier and James Waldo, 'AI Can Thrive in Open Societies', *Foreign Policy*, 13 June 2019.

74 Eliza Strickland, 'How IBM Watson Overpromised and Underdelivered on AI Health Care', *IEEE Spectrum*, 2 April 2019.

75 Martin U. Müller, 'Medical Applications Expose Current Limits of AI', *Spiegel*, 3 August 2018.

76 Angela Chen, 'IBM's Watson Gave Unsafe Recommendations For Treating Cancer', *Verge*, 26 July 2018.

77 Hal Hodson, 'Revealed: Google AI Has Access to Huge Haul of NHS Patient Data', *New Scientist*, 29 April 2016.

78 The ICO found that the Royal Free-DeepMind trial failed to comply with data protection law: https://ico.org.uk/about-the-ico/news-and-events/news-and-blogs/2017/07/royal-free-google-deepmind-trial-failed-to-comply-with-data-protection-law/

79 Julia Powles, 'DeepMind's Latest AI Health Breakthrough Has Some Problems', *Medium*, 6 August 2019.

80 Xiaoxuan Liu, Livia Faes, Aditya U. Kale, Siegfried K. Wagner, Dun Jack Fu, Alice Bruynseels, Thushika Mahendiran, Gabriella Moraes, Mohith Shamdas, Christoph

Kern, Joseph R. Ledsam, Martin K. Schmid, Konstantinos Balaskas, Eric J. Topol, Lucas M. Machmann, Pearse A. Keane and Alastair K. Denniston, 'A Comparison of Deep Learning Performance Against Health-Care Professionals in Detecting Diseases From Medical Imaging: A Systematic Review and Meta-Analysis', *Lancet Digital Health* 1, 2019.

81 L. Wang, L. Ding, Z. Liu, L. Sun, L. Chen, R. Jia, X. Dai, J. Cao and J. Ye, 'Automated Identification of Malignancy in Whole-Slide Pathological Images: Identification of Eyelid Malignant Melanoma in Gigapixel Pathological Slides Using Deep Learning', *British Journal of Ophthalmology* 104, 2020.

82 Margi Murphy, 'Privacy Concerns as Google Absorbs DeepMind's Health Division', *Telegraph*, 13 November 2018.

83 Julia Powles and Hal Hodson, 'Google DeepMind and Healthcare in an Age of Algorithms'.

84 Anne Trafton, 'Artificial Intelligence Yields New Antibiotic', MIT News Office, 20 February 2020.

85 In July 2020, the *New York Times* reported that even though Apple and Google had promised privacy to users of contact-tracing apps supported by their API, for the apps to work on Androids, users must turn on the device location setting, which enables GPS, even though the apps themselves only use Bluetooth. GPS is more privacy-invasive than Bluetooth because it records location. The fear is that Google might take advantage of the situation to collect and monetize users' location data. Natasha Singer, 'Google Promises Privacy With Virus App but Can Still Collect Location Data', *New York Times*, 20 July 2020. In July, Google updated the API. People with Android 11 can use the apps without turning on the device location setting, but turning on the setting for phones on older OS versions is still necessary. https://blog.google/inside-google/company-announcements/update-exposure-notifications/

86 Lorenzo Tondo, 'Scientists Say Mass Tests in Italian Town Have Halted Covid-19 There', *Guardian*, 18 March 2020.

87 'Covid-19: China's Qingdao to Test Nine Million in Five Days', BBC News, 12 October 2020.

88 Yves-Alexandre de Montjoye and his team have written a blog post about what they think are the biggest risks of coronavirus apps. Yves-Alexandre de Montjoye, Florimond Houssiau, Andrea Gadotti and Florent Guepin, 'Evaluating COVID-19 Contact Tracing Apps? Here Are 8 Privacy Questions We Think You Should Ask', Computational Privacy Group, 2 April 2020 (https://cpg.doc.ic.ac.uk/blog/evaluating-contact-tracing-apps-here-are-8-privacy-questions-we-think-you-should-ask/).

89 https://www.youtube.com/watch?v=_mzcbXi1Tkk

90 Naomi Klein, *The Shock Doctrine* (Random House, 2007).

91 Paul Mozur, Raymond Zhong and Aaron Krolik, 'In Coronavirus Fight, China Gives Citizens a Color Code, With Red Flags', *New York Times*, 1 March 2020.

92 Shanti Das and Shingi Mararike, 'Contact-Tracing Data Harvested From Pubs and Restaurants Being Sold On', *Times*, 11 October 2020.

93 Naomi Klein, 'Screen New Deal', *Intercept*, 8 May 2020.

94 Naomi Klein, 'Screen New Deal'.

95 Oscar Williams, 'Palantir's NHS Data Project "may outlive coronavirus crisis"', *New Statesman*, 30 April 2020.

96 Nick Statt, 'Peter Thiel's Controversial Palantir Is Helping Build a Coronavirus Tracking Tool for the Trump Admin', *Verge*, 21 April 2020.

97 Amy Thomson and Jonathan Browning, 'Peter Thiel's Palantir Is Given Access to U.K. Health Data on Covid-19 Patients', *Bloomberg*, 5 June 2020.

98 João Carlos Magalhães and Nick Couldry, 'Tech Giants Are Using This Crisis to Colonize the Welfare System', *Jacobin*, 27 April 2020.

99 Jon Henley and Robert Booth, 'Welfare Surveillance System Violates Human Rights, Dutch Court Rules', *Guardian*, 5 February 2020.

100 Yuval Harari, 'The World After Coronavirus', *Financial Times*, 20 March 2020.

101 'Big Tech's \$2trn Bull Run', *The Economist*, 22 February 2020.

CHAPTER SIX

1 Carissa Véliz, 'Why You Might Want to Think Twice About Surrendering Online Privacy for the Sake of Convenience', *The Conversation*, 11 January 2017.

2 Chris Wood, 'WhatsApp Photo Drug Dealer Caught By "Groundbreaking" Work', BBC News, 15 April 2018; Zoe Kleinman, 'Politician's Fingerprint "Cloned From Photos" By Hacker', BBC News, 29 December 2014.

3 Leo Kelion, 'Google Chief: I'd Disclose Smart Speakers Before Guests Enter My Home', BBC News, 15 October 2019.

4 In the Netherlands, a court has ordered a grandmother to delete all photos of her grandchildren that she posted on Facebook without their parents' permission. 'Grandmother Ordered to Delete Facebook Photos Under GDPR', BBC News, 21 May 2020.

5 Sonia Bokhari, 'I'm 14, and I Quit Social Media After Discovering What Was Posted About Me', *Fast Company*, 18 March 2019.

6 Sara Salinas, 'Six Top US Intelligence Chiefs Caution Against Buying Huawei Phones', CNBC, 13 February 2018.

7 Julien Gamba, Mohammed Rashed, Abbas Razaghpanah, Juan Tapiador and Narseo Vallina-Rodriguez, 'An Analysis of Pre-Installed Android Software', 41st IEEE Symposium on Security and Privacy, 2019.

8 Parmy Olson, 'Exclusive: WhatsApp Cofounder Brian Acton Gives the Inside Story On #DeleteFacebook and Why He Left \$850 Million Behind', *Forbes*, 26 September 2018.

9 William Turton, 'Why You Should Stop Using Telegram Right Now', *Gizmodo*, 24 June 2016.

10 Thanks to Ian Preston for letting me know about this trick.

11 https://blog.mozilla.org/security/2020/02/06/multi-account-containers-sync/

12 TJ McCue, '47 Percent of Consumers Are Blocking Ads', *Forbes*, 19 March 2019.

13 Christopher Wylie, *Mindf*ck. Inside Cambridge Analytica's Plot to Break the World*, 114.

14 Kim Zetter, 'The NSA Is Targeting Users of Privacy Services, Leaked Code Shows', *Wired*, 3 July 2014.

15 Kate O'Flaherty, 'Facebook Shuts Its Onavo Snooping App – But It Will Continue to Abuse User Privacy', *Forbes*, 22 February 2019.

16 Here are some guides to get you started, but you might want to check whether there are more up-to-date ones online: 'The Default Privacy Settings You Should Change and How to Do It', *Medium*, 18 July 2018; J. R. Raphael, '7 Google Privacy Settings You Should Revisit Right Now', *Fast Company*, 17 May 2019; Preston Gralla, 'How to Protect Your Privacy on Facebook', *Verge*, 7 June 2019.

17 Alex Hern, 'Are You A "Cyberhoarder"? Five Ways to Declutter Your Digital Life – From Emails to Photos', *Guardian*, 10 October 2018.

18 K. G. Orphanides, 'How to Securely Wipe Anything From Your Android, iPhone or PC', *Wired*, 26 January 2020.

19 For a list of the 10,000 most common passwords that you should avoid, see https://en.wikipedia.org/wiki/Wikipedia:10,000_most_common_passwords.

20 Finn Brunton and Helen Nissenbaum, *Obfuscation. A User's Guide for Privacy and Protest* (Cambridge MA: MIT Press, 2015), 1.

21 Alfred Ng, 'Teens Have Figured Out How to Mess With Instagram's Tracking Algorithm', CNET, 4 February 2020.

22 Hilary Osborne, 'Smart Appliances May Not be Worth Money in the Long Run, Warns Which?', *Guardian*, 8 June 2020.

23 Edwin Black, *IBM and the Holocaust*.

24 In reference to the Rohingya persecution, already Facebook has admitted to not 'doing enough to help prevent' the platform 'from being used to foment division and incite offline violence'. A UN fact-finding mission to Myanmar singled out Facebook as a 'useful instrument for those seeking to spread hate'. Many thousands of people have been killed. Hannah Ellis-Petersen, 'Facebook Admits Failings Over Incitement to Violence in Myanmar', *Guardian*, 6 November 2018.

25 Jack Poulson, 'I Used to Work for Google: I Am a Conscientious Objector', *New York Times*, 23 April 2019.

26 In the UK, Digital Catapult offers this kind of service. Full disclosure: I'm currently a member of its Ethics Committee.

27 Anna Wiener, 'Taking Back Our Privacy', *New Yorker*, 19 October 2020.

28 If Apple implements the privacy protections it has promised, Facebook will be under pressure to change some of its privacy-invasive practices. Megan Graham, 'Facebook Revenue Chief Says Ad-Supported Model Is "Under Assault" Amid Apple Privacy Changes', CNBC, 6 October 2020.

29 Andy Greenberg, 'A Guide to Getting Past Customs With Your Digital Privacy Intact', *Wired*, 12 February 2017.

30 Stéphane Hessel, *The Power of Indignation* (Skyhorse Publishing, 2012).

31 'The Data Economy. Special Report', *The Economist*, 20 February 2020.

CONCLUSION

1 Yuval Harari, 'The World After Coronavirus'.

REFERENCES

Abramowitz, Michael J., 'Democracy in Crisis', *Freedom in the World* (2018)

Ajunwa, Ifeoma, Kate Crawford and Jason Schultz, 'Limitless Worker Surveillance', *California Law Review* 105, 2017

Allard, Jody, 'How Gene Testing Forced Me to Reveal My Private Health Information', *Vice*, 27 May 2016

Ambrose, Jillian, 'Lights Stay On Despite Cyber-Attack on UK's Electricity System', *Guardian*, 14 May 2020

Angwin, Julia, *Dragnet Nation* (New York: Times Books, 2014)

Angwin, Julia, Jeff Larson, Charlie Savage, James Risen, Henrik Moltke and Laura Poitras, 'NSA Spying Relies on AT&T's "Extreme Willingness to Help"', *ProPublica*, 15 August 2015

Associated Press, 'Google Records Your Location Even When You Tell It Not To', *Guardian*, 13 August 2018

Balkin, Jack M., 'Information Fiduciaries and the First Amendment', *UC Davis Law Review* 49, 2016

Bamford, Roxanne, Benedict Macon-Cooney, Hermione Dace and Chris Yiu, 'A Price Worth Paying: Tech, Privacy and the Fight Against Covid-19' (Tony Blair Institute for Global Change, 2020)

Baraniuk, Chris, 'Ashley Madison: "Suicides" Over Website Hack', BBC News, 24 August 2015

Bassett, Laura, 'Digital Media Is Suffocating – and It's Facebook and Google's Fault', *American Prospect*, 6 May 2019

Battelle, John, 'The Birth of Google', *Wired*, 1 August 2005

Baxter, Michael, 'Do Connected Cars Pose a Privacy Threat?', *GDPR: Report*, 1 August 2018

Beckett, Lois, 'Everything We Know About What Data Brokers Know About You', *ProPublica*, 13 June 2014

Beckett, Lois, 'Under Digital Surveillance: How American Schools Spy on Millions of Kids', *Guardian*, 22 October 2019

Bell, Emily, 'Why Facebook's News Feed Changes Are Bad News for Democracy', *Guardian*, 21 January 2018

Bharat, Krishna, Stephen Lawrence and Meham Sahami, 'Generating User Information for Use in Targeted Advertising' (2003)

Biba, Erin, 'How Connected Car Tech Is Eroding Personal Privacy', BBC News, 9 August 2016

Biddle, Sam, 'For Owners of Amazon's Ring Security Cameras, Strangers May Have Been Watching Too', *Intercept*, 10 January 2019

Biddle, Sam, 'How Peter Thiel's Palantir Helped the NSA Spy on the Whole World', *Intercept*, 22 February 2017

Biddle, Sam, 'In Court, Facebook Blames Users for Destroying Their Own Right to Privacy', *Intercept*, 14 June 2014

'Big Tech's $2trn Bull Run', *The Economist*, 22 February 2020

Bilton, Nick, 'Why Google Glass Broke', *New York Times*, 4 February 2015

Black, Edwin, *IBM and the Holocaust* (Washington, DC: Dialog Press, 2012)

Bokhari, Sonia, 'I'm 14, and I Quit Social Media After Discovering What Was Posted About Me', *Fast Company*, 18 March 2019

Bond, Robert M. et al, 'A 61-Million-Person Experiment in Social Influence and Political Mobilization', *Nature* 489, 2012

Booth, Robert, Sandra Laville and Shiv Malik, 'Royal Wedding: Police Criticised for Pre-Emptive Strikes Against Protestors', *Guardian*, 29 April 2011

Brin, Sergey and Lawrence Page, 'The Anatomy of a Large-Scale Hypertextual Web Search Engine', *Computer Networks and ISDN Systems* 30, 1998

'British Airways Faces Record £183m Fine for Data Breach', BBC News, 8 July 2019

Brooke, Siân and Carissa Véliz, 'Views on Privacy. A Survey', *Data, Privacy & the Individual* (Center for the Governance of Change, IE University, 2020)

Brown, Kristen V., 'What DNA Testing Companies' Terrifying Privacy Policies Actually Mean', *Gizmodo*, 18 October 2017

Brunton, Finn and Helen Nissenbaum, *Obfuscation. A User's Guide for Privacy and Protest* (Cambridge MA: MIT Press, 2015)

Bryant, Ben, 'VICE News Investigation Finds Signs of Secret Phone Surveillance Across London', *Vice*, 15 January 2016

Burgess, Matt, 'More Than 1,000 UK Schools Found To Be Monitoring Children With Surveillance Software', *Wired*, 8 November 2016

Burr, Christopher and Nello Cristianini, 'Can Machines Read our Minds?', *Minds and Machines* 29, 2019

Carr, Austin, Matt Day, Sarah Frier and Mark Gurman, 'Silicon Valley Is Listening to Your Most Intimate Moments', *Bloomberg Businessweek*, 11 December 2019

Caruso, Jay, 'The Latest Battleground Poll Tells Us Democrats Are Over-Correcting for 2020 – and They Can't Beat Trump That Way', *Independent*, 5 November 2019

Chen, Angela, 'IBM's Watson Gave Unsafe Recommendations For Treating Cancer', *Verge*, 26 July 2018

Chen, Angela and Alessandra Potenza, 'Cambridge Analytica's Facebook Data Abuse Shouldn't Get Credit for Trump', *Verge*, 20 March 2018

Christman, John, 'Autonomy in Moral and Political Philosophy', in Edward N. Zalta (ed.), *The Stanford Encyclopedia of Philosophy* (2015)

Clifford, Stephanie and Quentin Hardy, 'Attention, Shoppers: Store Is Tracking Your Cell', *New York Times*, 14 July 2013

Cockburn, Harry, 'The UK's Strangest Laws That Are Still Enforced', *Independent*, 8 September 2016

Coldewey, Devin, 'Grindr Sends HIV Status to Third Parties, and Some Personal Data Unencrypted', *TechCrunch*, 2 April 2018

Cole, David, ' "We Kill People Based on Metadata" ', *New York Review of Books*, 10 May 2014

Cox, Joseph, 'I Gave a Bounty Hunter $300. Then He Located Our Phone', *Motherboard*, 8 January 2019

Cox, Joseph, 'Revealed: Microsoft Contractors Are Listening to Some Skype Calls', *Motherboard*, 7 August 2019

Criado Perez, Caroline, *Invisible Women. Exposing Data Bias in a World Designed for Men* (Vintage, 2019)

Curran, Dylan, 'Are You Ready? Here Is All The Data Facebook And Google Have On You', *Guardian*, 30 March 2018

Dance, Gabriel J. X., Michael LaForgia and Nicholas Confessore, 'As Facebook Raised a Privacy Wall, It Carved an Opening for Tech Giants', *New York Times*, 18 December 2018

Daniel, Caroline and Maija Palmer, 'Google's Goal: To Organise Your Daily Life', *Financial Times*, 22 May 2007

'The Data Economy. Special Report', *The Economist*, 20 February 2020

Davies, Jessica, 'After GDPR, The New York Times Cut Off Ad Exchanges in Europe – and Kept Growing Ad Revenue', *Digiday*, 16 January 2019

de Montjoye, Y. A., C. A. Hidalgo, M. Verleysen and V. D. Blondel, 'Unique in the Crowd: The Privacy Bounds of Human Mobility', *Scientific Reports* 3, 2013

de Montjoye, Y. A., L. Radaelli, V. K. Singh and A. S. Pentland, 'Identity and privacy. Unique in the Shopping Mall: On the Reidentifiability of Credit Card Metadata', *Science* 347, 2015

de Zwart, Hans, 'During World War II, We Did Have Something to Hide', *Medium*, 30 April 2015

'The Default Privacy Settings You Should Change and How to Do It', *Medium*, 18 July 2018

Digital, Culture, Media and Sport Committee, 'Disinformation and "Fake News": Final Report' (House of Commons, 2019)

Douglas, Thomas and Lauren Van den Borre, 'Asbestos neglect: Why asbestos exposure deserves greater policy attention', *Health Policy* 123, 2019

Dreyfus, Hubert and Paul Rabinow, *Michel Foucault. Beyond Structuralism and Hermeneutics* (University of Chicago Press, 1982)

Dubois, Daniel J., Roman Kolcun, Anna Maria Mandalari, Muhammad Talha Paracha, David Choffnes and Hamed Haddadi, 'When Speakers Are All Ears', *Proceedings on 20th Privacy Enhancing Technologies Symposium*, 2020

Dunn, Will, 'Can Nuclear Weapons Be Hacked?', *New Statesman*, 7 May 2018

Dunne, Carey, 'Ten easy encryption tips for warding off hackers, the US government – and Russia', *Quartz*, 4 January 2017

Dwoskin, Elizabeth, 'FTC: Data Brokers Can Buy Your Bank Account Number for 50 Cents', *Wall Street Journal*, 24 December 2014

Dwoskin, Elizabeth and Tony Romm, 'Facebook's Rules for Accessing User Data Lured More Than Just Cambridge Analytica', *Washington Post*, 20 March 2018

'Economic Impact of Advertising in the United States' (IHS Economics and Country Risk, 2015)

The Economist Intelligence Unit, 'Democracy Index 2019. A Year of Democratic Setbacks and Popular Protest' (2019)

Edwards, Douglas, *I'm Feeling Lucky: The Confessions of Google Employee Number 59* (Houghton Mifflin Harcourt, 2011)

Ellis-Petersen, Hannah, 'Facebook Admits Failings Over Incitement to Violence in Myanmar', *Guardian*, 6 November 2018

Englehardt, Steven, Jeffrey Han and Arvind Narayanan, 'I Never Signed Up For This! Privacy Implications of Email Tracking', *Proceedings on Privacy Enhancing Technologies* 1, 2018

Esguerra, Richard, 'Google CEO Eric Schmidt Dismisses the Importance of Privacy', *Electronic Frontier Foundation*, 10 December 2009

Eveleth, Rose, 'The Biggest Lie Tech People Tell Themselves – and the Rest of Us', *Vox*, 8 October 2019

'Facebook Fined £500,000 for Cambridge Analytica Scandal', BBC News, 25 October 2018

Flyvbjerg, Bent, *Rationality and Power. Democracy in Practice* (Chicago University Press, 1998)

Foroohar, Rana, 'Year in a Word: Techlash', *Financial Times*, 16 December 2018

Forst, Rainer, 'Noumenal Power', *Journal of Political Philosophy* 23, 2015

Foucault, Michel, *Discipline and Punish* (London: Penguin Books, 1977)

Fowler, Geoffrey, 'The Doorbells Have Eyes: The Privacy Battle Brewing Over Home Security Cameras', *Washington Post*, 31 January 2019

Franceschi-Bicchierai, Lorenzo, 'Russian Facebook Trolls Got Two Groups of People to Protest Each Other in Texas', *Motherboard*, 1 November 2017

Frederik, Jesse and Maurits Martijn, 'The New Dot Com Bubble Is Here: It's Called Online Advertising', *Correspondent*, 6 November 2019

Frey, Chris, 'Revealed: How Facial Recognition Has Invaded Shops – and Your Privacy', *Guardian*, 3 March 2016

'FTC Imposes $5 Billion Penalty and Sweeping New Privacy Restrictions on Facebook', FTC Press Release, 24 July 2019

Fung, Brian, 'How Stores Use Your Phone's WiFi to Track Your Shopping Habits', *Washington Post*, 19 October 2013

Galdon Clavell, Gemma, 'Protect Rights at Automated Borders', *Nature* 543, 2017

Gamba, Julien, Mohammed Rashed, Abbas Razaghpanah, Juan Tapiador and Narseo Vallina-Rodriguez, 'An Analysis of Pre-Installed Android Software', 41st IEEE Symposium on Security and Privacy, 2019

Gan, Nectar, 'China Is Installing Surveillance Cameras Outside People's Front Doors . . . and Sometimes Inside Their Homes', *CNN Business*, 28 April 2020

Gibbs, Samuel and Alex Hern, 'Google at 20: How Two "Obnoxious" Students Changed the Internet', *Guardian*, 24 September 2018

Glanz, James and Andrew W. Lehren, 'NSA Spied on Allies, Aid Groups and Businesses', *New York Times*, 21 December 2013

'The Government Uses "Near Perfect Surveillance" Data on Americans', *New York Times*, 7 February 2020

Graham, Richard, 'Google and advertising: digital capitalism in the context of Post-Fordism, the reification of language, and the rise of fake news', *Palgrave Communications* 3, 2017

Gralla, Preston, 'How to Protect Your Privacy on Facebook', *Verge*, 7 June 2019

Gramlich, John, '10 Facts About Americans and Facebook', Pew Research Center, 16 May 2019

'Grandmother Ordered to Delete Facebook Photos Under GDPR', BBC News, 21 May 2020

Grassegger, Hannes, 'Facebook Says Its "Voter Button" Is Good for Turnout. But Should the Tech Giant Be Nudging Us at All?', *Observer*, 15 April 2018

Grauer, Yael, 'What Are "Data Brokers," and Why Are They Scooping Up Information About You?', *Motherboard*, 27 May 2018

Greenberg, Andy, 'A Guide to Getting Past Customs With Your Digital Privacy Intact', *Wired*, 12 February 2017

Greenberg, Andy, 'How Hacked Water Heaters Could Trigger Mass Blackouts', *Wired*, 13 August 2018

Greenberg, Andy, 'New Clues Show How Russia's Grid Hackers Aimed for Physical Destruction', *Wired*, 12 September 2019

Grothaus, Michael, 'Forget the New iPhones: Apple's Best Product Is Now Privacy', *Fast Company*, 13 September 2018

Guimón, Pablo, ' "Brexit Wouldn't Have Happened Without Cambridge Analytica" ',
  El País, 27 March 2018

Hagey, Keach, 'Behavioral Ad Targeting Not Paying Off for Publishers, Study
  Suggests', Wall Street Journal, 29 May 2019

Halpern, Sue, 'Cambridge Analytica and the Perils of Psychographics', New Yorker,
  30 March 2018

Hambling, David, 'The Pentagon Has a Laser That Can Identify People From a
  Distance – By Their Heartbeat', MIT Technology Review, 27 June 2019

Harari, Yuval, 'The World After Coronavirus', Financial Times, 20 March 2020

Hartzog, Woodrow and Evan Selinger, 'Facial Recognition Is the Perfect Tool for
  Oppression', Medium, 2 August 2018

Harvey, Fiona, 'Ozone Layer Finally Healing After Damage Caused by Aerosols, UN
  Says', Guardian, 5 November 2018

Heaven, Douglas, 'An AI Lie Detector Will Interrogate Travellers at Some EU
  Borders', New Scientist, 31 October 2018

Helm, Toby, 'Patient Data From GP Surgeries Sold to US Companies', Observer,
  7 December 2019

Henley, Jon and Robert Booth, 'Welfare Surveillance System Violates Human Rights,
  Dutch Court Rules', Guardian, 5 February 2020

Hern, Alex, 'Apple Contractors "Regularly Hear Confidential Details" on Siri
  Recordings', Guardian, 26 July 2019

Hern, Alex, 'Apple Whistleblower Goes Public Over "Lack of Action" ', Guardian,
  20 May 2020

Hern, Alex, 'Are You A "Cyberhoarder"? Five Ways to Declutter Your Digital
  Life – From Emails to Photos', Guardian, 10 October 2018

Hern, Alex, 'Facebook Admits Contractors Listened to Users' Recordings Without
  Their Knowledge', Guardian, 14 August 2019

Hern, Alex, 'Facebook "Dark Ads" Can Swing Political Opinions, Research Shows',
  Guardian, 31 July 2017

Hern, Alex, 'Facebook Faces Backlash Over Users' Safety Phone Numbers', Guardian,
  4 March 2019

Hern, Alex, 'Hackers Publish Private Photos From Cosmetic Surgery Clinic',
  Guardian, 31 May 2017

Hern, Alex, 'Netflix's Biggest Competitor? Sleep', Guardian, 18 April 2017

Hern, Alex, 'Privacy Policies of Tech Giants "Still Not GDPR-Compliant" ', Guardian,
  5 July 2018

Hern, Alex, 'Smart Electricity Meters Can Be Dangerously Insecure, Warns Expert',
  Guardian, 29 December 2016

Hern, Alex, 'UK Homes Vulnerable to "Staggering" Level of Corporate Surveillance',
  Guardian, 1 June 2018

Hernández, José Antonio, 'Me han robado la identidad y estoy a base de lexatín; yo
  no soy una delincuente', El País, 24 August 2016

Hessel, Stéphane, *The Power of Indignation* (Skyhorse Publishing, 2012)

Hicken, Melanie, 'Data Brokers Selling Lists of Rape Victims, AIDS Patients', CNN, 19 December 2013

Hill, Kashmir, 'Facebook Added "Research" To User Agreement 4 Months After Emotion Manipulation Study', *Forbes*, 30 June 2014

Hill, Kashmir, 'Facebook Recommended That This Psychiatrist's Patients Friend Each Other', *Splinter News*, 29 August 2016

Hill, Kashmir, 'Facebook Was Fully Aware That Tracking Who People Call and Text Is Creepy But Did It Anyway', *Gizmodo*, 12 May 2018

Hill, Kashmir, 'How Facebook Outs Sex Workers', *Gizmodo*, 10 November 2017

Hill, Kashmir, 'I Got Access to My Secret Consumer Score. Now You Can Get Yours, Too', *New York Times*, 4 November 2019

Hill, Kashmir, '"People You May Know": A Controversial Facebook Feature's 10-Year History', *Gizmodo*, 8 August 2018

Hill, Kashmir and Aaron Krolik, 'How Photos of Your Kids Are Powering Surveillance Technology', *New York Times*, 11 October 2019

Hodson, Hal, 'Revealed: Google AI Has Access to Huge Haul of NHS Patient Data', *New Scientist*, 29 April 2016

Hoffman, Anna Lauren, 'Facebook is Worried About Users Sharing Less – But it Only Has Itself to Blame', *Guardian*, 19 April 2016

Hoffman, David A., 'Intel Executive: Rein In Data Brokers', *New York Times*, 15 July 2019

Holpuch, Amanda, 'Trump's Separation of Families Constitutes Torture, Doctors Find', *Guardian*, 25 February 2020

Hsu, Jeremy, 'The Strava Heat Map and the End of Secrets', *Wired*, 29 January 2018

Hsu, Tiffany, 'The Advertising Industry Has a Problem: People Hate Ads', *New York Times*, 28 October 2019

Isikoff, Michael, 'NSA Program Stopped No Terror Attacks, Says White House Panel Member', NBC News, 19 December 2013

Jenkins, Holman W., 'Google and the Search for the Future', *Wall Street Journal*, 14 August 2010

Johnson, Bobbie, 'Facebook Privacy Change Angers Campaigners', *Guardian*, 10 December 2009

Johnson, Bobbie, 'Privacy No Longer a Social Norm, Says Facebook Founder', *Guardian*, 11 January 2010

Johnston, Casey, 'Facebook Is Tracking Your "Self-Censorship"', *Wired*, 17 December 2013

Jones, Rupert, 'Identity Fraud Reaching Epidemic Levels, New Figures Show', *Guardian*, 23 August 2017

Kaiser, Brittany, *Targeted. My Inside Story of Cambridge Analytica and How Trump, Brexit and Facebook Broke Democracy* (HarperCollins, 2019)

Kaiser, Jocelyn, 'We Will Find You: DNA Search Used to Nab Golden State Killer Can Home In On About 60% of White Americans', *Science Magazine*, 11 October 2018

Kang, Cecilia and Mike Isaac, 'Defiant Zuckerberg Says Facebook Won't Police Political Speech', *New York Times*, 17 October 2019

Kang, Cecilia and Kenneth P. Vogel, 'Tech Giants Amass a Lobbying Army for an Epic Washington Battle', *New York Times*, 5 June 2019

Kayyem, Juliette, 'Never Say "Never Again"', *Foreign Policy*, 11 September 2012

Kelion, Leo, 'Google Chief: I'd Disclose Smart Speakers Before Guests Enter My Home', BBC News, 15 October 2019

Khan, Lina and David E. Pozen, 'A Skeptical View of Information Fiduciaries', *Harvard Law Review* 133, 2019

Khandaker, Tamara, 'Canada Is Using Ancestry DNA Websites To Help It Deport People', *Vice*, 26 July 2018

Kim, Tae, 'Warren Buffett Believes This Is "the Most Important Thing" to Find in a Business', CNBC, 7 May 2018

Kim, Tami, Kate Barasz and Leslie K. John, 'Why Am I Seeing This Ad? The Effect of Ad Transparency on Ad Effectiveness', *Journal of Consumer Research* 45, 2019

Klein, Naomi, 'Screen New Deal', *Intercept*, 8 May 2020

Klein, Naomi, *The Shock Doctrine* (Random House, 2007)

Kleinman, Zoe, 'Politician's Fingerprint "Cloned From Photos" By Hacker', BBC News, 29 December 2014

Knoema, 'United States of America – Contribution of Travel and Tourism to GDP as a Share of GDP' (2018)

Kobie, Nicole, 'Heathrow's Facial Recognition Tech Could Make Airports More Bearable', *Wired*, 18 October 2018

Koepke, Logan, '"We Can Change These Terms at Anytime": The Detritus of Terms of Service Agreements', *Medium*, 18 January 2015

Koerner, Brendan I., 'Your Relative's DNA Could Turn You Into a Suspect', *Wired*, 13 October 2015

Kosinski, Michal, David Stillwell and Thore Graepel, 'Private Traits and Attributes Are Predictable From Digital Records of Human Behavior', *PNAS* 110, 2013

Kramer, Alexis, 'Forced Phone Fingerprint Swipes Raise Fifth Amendment Questions', *Bloomberg Law*, 7 October 2019

Kurra, Babu, 'How 9/11 Completely Changed Surveillance in U.S.', *Wired*, 11 September 2011

Lamont, Tom, 'Life After the Ashley Madison Affair', *Observer*, 27 February 2016

Lapowsky, Issie, 'The 21 (and Counting) Biggest Facebook Scandals of 2018', *Wired*, 20 December 2018

Lecher, Colin, 'Strava Fitness App Quietly Added a New Opt-Out for Controversial Heat Map', *Verge*, 1 March 2018

Lee, Jennifer, 'Postcards From Planet Google', *New York Times*, 28 November 2002

Lee, Micah and Yael Grauer, 'Zoom Meetings Aren't End-to-End Encrypted, Despite Misleading Marketing', *Intercept*, 31 March 2020

Levin, Sam, 'Tech Firms Make Millions from Trump's Anti-Immigrant Agenda, Report Finds', *Guardian*, 23 October 2018

Levitsky, Steven and Daniel Ziblatt, *How Democracies Die* (Penguin, 2018)

Levy, Steven, *In the Plex. How Google Thinks, Works, and Shapes Our Lives* (New York: Simon & Schuster, 2011)

Liu, Xiaoxuan, Livia Faes, Aditya U. Kale, Siegfried K. Wagner, Dun Jack Fu, Alice Bruynseels, Thushika Mahendiran, Gabriella Moraes, Mohith Shamdas, Christoph Kern, Joseph R. Ledsam, Martin K. Schmid, Konstantinos Balaskas, Eric J. Topol, Lucas M. Machmann, Pearse A. Keane and Alastair K. Denniston, 'A Comparison of Deep Learning Performance Against Health-Care Professionals in Detecting Diseases From Medical Imaging: A Systematic Review and Meta-Analysis', *Lancet Digital Health* 1, 2019

Lomas, Natasha, 'A Brief History of Facebook's Privacy Hostility Ahead of Zuckerberg's Testimony', *TechCrunch*, 10 April 2018

Lomas, Natasha, 'The Case Against Behavioral Advertising Is Stacking Up', *TechCrunch*, 20 January 2019

Louis, Tristan, 'How Much Is a User Worth?', *Forbes*, 31 August 2013

Lukes, Steven, *Power. A Radical View* (Red Globe Press, 2005)

Lyngaas, Sean, 'Hacking Nuclear Systems Is the Ultimate Cyber Threat. Are We Prepared?', *Verge*, 23 January 2018

Macintyre, Amber, 'Who's Working for Your Vote?', *Tactical Tech*, 29 November 2018

MacLachlan, Alice, 'Fiduciary Duties and the Ethics of Public Apology', *Journal of Applied Philosophy* 35, 2018

Magalhães, João Carlos and Nick Couldry, 'Tech Giants Are Using This Crisis to Colonize the Welfare System', *Jacobin*, 27 April 2020

Mahdawi, Arwa, 'Spotify Can Tell If You're Sad. Here's Why That Should Scare You', *Guardian*, 16 September 2018

Malin, Bradley and Latanya Sweeney, 'Determining the Identifiability of DNA Database Entries', *Proceedings, Journal of the American Medical Informatics Association*, 2000

'A Manifesto for Renewing Liberalism', *The Economist*, 15 September 2018

Marcus, Gary, 'Total Recall: The Woman Who Can't Forget', *Wired*, 23 March 2009

Matsakis, Louise, 'Online Ad Targeting Does Work – As Long As It's Not Creepy', *Wired*, 11 May 2018

Matsakis, Louise, 'The WIRED Guide to Your Personal Data (and Who Is Using It)', *Wired*, 15 February 2019

Maxmen, Amy, 'Surveillance Science', *Nature* 569, 2019

Mayer-Schönberger, Viktor, *Delete. The Virtue of Forgetting in the Digital Age* (Princeton University Press, 2009)

McCue, TJ, '47 Percent of Consumers Are Blocking Ads', *Forbes*, 19 March 2019

Merchant, Brian, 'How Email Open Tracking Quietly Took Over the Web', *Wired*, 11 December 2017

Metz, Rachel, 'The Smartphone App That Can Tell You're Depressed Before You Know it Yourself', *MIT Technology Review*, 15 October 2018

Michel, Chloé, Michelle Sovinsky, Eugenio Proto and Andrew Oswald, 'Advertising as a Major Source of Human Dissatisfaction: Cross-National Evidence on One Million Europeans', in M. Rojas (ed.), *The Economics of Happiness* (Springer, 2019)

Midgley, Clare, 'Slave Sugar Boycotts, Female Activism and the Domestic Base of British Anti-Slavery Culture', *Slavery and Abolition* 17, 1996

Miles, Tom, 'UN Surveillance Expert Urges Global Moratorium on Sale of Spyware', Reuters, 18 June 2019

Mill, John Stuart, *Collected Works of John Stuart Mill* (University of Toronto Press, 1963)

Mill, John Stuart, *On Liberty* (Indianapolis: Hackett Publishing Company, 1978)

Mims, Christopher, 'Here Comes "Smart Dust," The Tiny Computers That Pull Power From The Air', *Wall Street Journal*, 8 November 2018

Mistreanu, Simina, 'Life Inside China's Social Credit Laboratory', *Foreign Policy*, 3 April 2018

Molla, Rani, 'These Publications Have the Most to Lose From Facebook's New Algorithm Changes', *Vox*, 25 January 2018

Moore, Barrington, *Privacy. Studies in Social and Cultural History* (Armonk, New York: M. E. Sharpe, 1984)

Mozur, Paul, Raymond Zhong and Aaron Krolik, 'In Coronavirus Fight, China Gives Citizens a Color Code, With Red Flags', *New York Times*, 1 March 2020

Müller, Martin U., 'Medical Applications Expose Current Limits of AI', *Spiegel*, 3 August 2018

Munro, Dan, 'Data Breaches In Healthcare Totaled Over 112 Million Records in 2015', *Forbes*, 31 December 2015

Murphy, Erin E., *Inside the Cell. The Dark Side of Forensic DNA* (Nation Books, 2015)

Murphy, Margi, 'Privacy concerns as Google absorbs DeepMind's health division', *Telegraph*, 13 November 2018

Nagel, Thomas, 'Concealment and Exposure', *Philosophy and Public Affairs* 27, 1998

Nakashima, Ellen and Joby Warrick, 'Stuxnet Was Work of US and Israeli Experts, Officials Say', *Washington Post*, 2 June 2012

'Nature's Language Is Being Hijacked By Technology', BBC News, 1 August 2019

Naughton, John, 'More Choice On Privacy Just Means More Chances To Do What's Best For Big Tech', *Guardian*, 8 July 2018

Neff, Gina and Dawn Nafus, *Self-Tracking* (MIT Press, 2016)

Newman, Lily Hay, 'How to Block the Ultrasonic Signals You Didn't Know Were Tracking You', *Wired*, 3 November 2016

Newton, Casey, 'How Grindr Became a National Security Issue', *Verge*, 28 March 2019

Ng, Alfred, 'Teens Have Figured Out How to Mess With Instagram's Tracking Algorithm', CNET, 4 February 2020

Ng, Alfred, 'With Smart Sneakers, Privacy Risks Take a Great Leap', CNET, 13 February 2019

Nguyen, Nicole, 'If You Have a Smart TV, Take a Closer Look at Your Privacy Settings', CNBC, 9 March 2017

Noble, Safiya, *Algorithms of Oppression. How Search Engines Reinforce Racism* (NYU Press, 2018)

O'Flaherty, Kate, 'Facebook Shuts Its Onavo Snooping App – But It Will Continue to Abuse User Privacy', *Forbes*, 22 February 2019

Ogilvy, David, *Confessions of an Advertising Man* (Harpenden: Southbank Publishing, 2013)

O'Hara, Kieron and Nigel Shadbolt, 'Privacy on the Data Web', *Communications of the ACM* 53, 2010

O'Harrow Jr, Robert, 'Online Firm Gave Victim's Data to Killer', *Chicago Tribune*, 6 January 2006

Oliver, Myrna, 'Legends Nureyev, Gillespie Die: Defector Was One of Century's Great Dancers', *Los Angeles Times*, 7 January 1993

Olson, Parmy, 'Exclusive: WhatsApp Cofounder Brian Acton Gives the Inside Story On #DeleteFacebook and Why He Left $850 Million Behind', *Forbes*, 26 September 2018

Orphanides, K. G., 'How to Securely Wipe Anything From Your Android, iPhone or PC', *Wired*, 26 January 2020

Orwell, George, *Fascism and Democracy* (Penguin, 2020)

Orwell, George, *Politics and the English Language* (Penguin, 2013)

Osborne, Hilary, 'Smart Appliances May Not be Worth Money in the Long Run, Warns Which?', *Guardian*, 8 June 2020

O'Sullivan, Donie and Drew Griffin, 'Cambridge Analytica Ran Voter Suppression Campaigns, Whistleblower Claims', CNN, 17 May 2018

Parcak, Sarah, 'Are We Ready for Satellites That See Our Every Move?', *New York Times*, 15 October 2019

Parkin, Simon, 'Has Dopamine Got Us Hooked on Tech?', *Guardian*, 4 March 2018

Paul, Kari, 'Zoom to Exclude Free Calls From End-to-End Encryption to Allow FBI Cooperation', *Guardian*, 4 June 2020

Penney, Jonathon W., 'Chilling Effects: Online Surveillance and Wikipedia Use', *Berkeley Technology Law Journal* 31, 2016

Pérez Colomé, Jordi, 'Por qué China roba datos privados de decenas de millones de estadounidenses', *El País*, 17 February 2020

Peterson, Andrea, 'Snowden filmmaker Laura Poitras: "Facebook is a gift to intelligence agencies"', *Washington Post*, 23 October 2014

Phillips, Dom, 'Brazil's Biggest Newspaper Pulls Content From Facebook After Algorithm Change', *Guardian*, 8 February 2018

Poole, Steven, 'Drones the Size of Bees – Good or Evil?', *Guardian*, 14 June 2013

Popper, Karl, *The Open Society and Its Enemies* (Routledge, 2002)

Poulson, Jack, 'I Used to Work for Google. I Am a Conscientious Objector', *New York Times*, 23 April 2019

Powles, Julia, 'DeepMind's Latest AI Health Breakthrough Has Some Problems', *Medium*, 6 August 2019

Powles, Julia and Enrique Chaparro, 'How Google Determined Our Right to be Forgotten', *Guardian*, 18 February 2015

Powles, Julia and Hal Hodson, 'Google DeepMind and Healthcare in an Age of Algorithms', *Health and Technology 7*, 2017

Price, Rob, 'An Ashley Madison User Received a Terrifying Blackmail Letter', *Business Insider*, 22 January 2016

'Privacy Online: Fair Information Practices in the Electronic Marketplace. A Report to Congress' (Federal Trade Commission, 2000)

Purdy, Jebediah, 'The Anti-Democratic Worldview of Steve Bannon and Peter Thiel', *Politico*, 30 November 2016

Quain, John R., 'Cars Suck Up Data About You. Where Does It All Go?', *New York Times*, 27 July 2017

Ralph, Oliver, 'Insurance and the Big Data Technology Revolution', *Financial Times*, 24 February 2017

Ram, Aliya and Emma Boyde, 'People Love Fitness Trackers, But Should Employers Give Them Out?', *Financial Times*, 16 April 2018

Ram, Aliya and Madhumita Murgia, 'Data Brokers: Regulators Try To Rein In The "Privacy Deathstars"', *Financial Times*, 8 January 2019

Ramsey, Lydia and Samantha Lee, 'Our DNA is 99.9% the Same as the Person Next to Us – and We're Surprisingly Similar to a Lot of Other Living Things', *Business Insider*, 3 April 2018

Raphael, J. R., '7 Google Privacy Settings You Should Revisit Right Now', *Fast Company*, 17 May 2019

'Report on the President's Surveillance Program' (2009)

Revell, Timothy, 'How to Turn Facebook Into a Weaponised AI Propaganda Machine', *New Scientist*, 28 July 2017

Rogers, Kaleigh, 'Let's Talk About Mark Zuckerberg's Claim That Facebook "Doesn't Sell Data"', *Motherboard*, 11 April 2018

Romm, Tony, 'Tech Giants Led by Amazon, Facebook and Google Spent Nearly Half a Billion on Lobbying Over the Last Decade', *Washington Post*, 22 January 2020

Russell, Bertrand, *Power. A New Social Analysis* (Routledge, 2004)

Salinas, Sara, 'Six Top US Intelligence Chiefs Caution Against Buying Huawei Phones', CNBC, 13 February 2018

Sanger, David E., 'Hackers Took Fingerprints of 5.6 Million U.S. Workers, Government Says', *New York Times*, 23 September 2015

Sanghani, Radhika, 'Your Boss Can Read Your Personal Emails. Here's What You Need To Know', *Telegraph*, 14 January 2016

Satariano, Adam, 'Europe's Privacy Law Hasn't Shown Its Teeth, Frustrating Advocates', *New York Times*, 27 April 2020

Savage, Charlie, 'Declassified Report Shows Doubts About Value of N.S.A.'s Warrantless Spying', *New York Times*, 25 April 2015

Savage, Charlie, *Power Wars. Inside Obama's Post-9/11 Presidency* (New York: Little, Brown and Company, 2015)

Schilit, S. L. and A. Schilit Nitenson, 'My Identical Twin Sequenced our Genome', *Journal of Genetic Counseling* 26, 2017

Schneier, Bruce, *Click Here to Kill Everybody. Security and Survival in a Hyper-Connected World* (New York: W. W. Norton & Company, 2018)

Schneier, Bruce, *Data and Goliath* (London: W. W. Norton & Company, 2015)

Schneier, Bruce, 'Data is a toxic asset, so why not throw it out?', CNN, 1 March 2016

Schneier, Bruce and James Waldo, 'AI Can Thrive in Open Societies', *Foreign Policy*, 13 June 2019

Selinger, Evan and Woodrow Hartzog, 'What Happens When Employers Can Read Your Facial Expressions?', *New York Times*, 17 October 2019

Seltzer, William and Margo Anderson, 'The Dark Side of Numbers: The Role of Population Data Systems in Human Rights Abuses', *Social Research* 68, 2001

Shaban, Hamza, 'Google for the First Time Outspent Every Other Company to Influence Washington in 2017', *Washington Post*, 23 January 2018

Shadbolt, Nigel and Roger Hampson, *The Digital Ape. How to Live (in Peace) with Smart Machines* (Oxford University Press, 2019)

Shaer, Matthew, 'The False Promise of DNA Testing', *Atlantic* (June 2016)

Sherman, Len, 'Zuckerberg's Broken Promises Show Facebook Is Not Your Friend', *Forbes*, 23 May 2018

Shontell, Alyson, 'Mark Zuckerberg Just Spent More Than $30 Million Buying 4 Neighboring Houses So He Could Have Privacy', *Business Insider*, 11 October 2013

Singel, Ryan, 'Netflix Spilled Your Brokeback Mountain Secret, Lawsuit Claims', *Wired*, 17 December 2009

Singer, Natasha, 'Data Broker Is Charged With Selling Consumers' Financial Details to "Fraudsters"', *New York Times*, 23 December 2014

Singer, Natasha, 'Facebook's Push For Facial Recognition Prompts Privacy Alarms', *New York Times*, 9 July 2018

Smith, Dave and Phil Chamberlain, 'On the Blacklist: How Did the UK's Top Building Firms Get Secret Information on Their Workers?', *Guardian*, 27 February 2015

Smith, David, 'How Key Republicans Inside Facebook Are Shifting Its Politics to the Right', *Guardian*, 3 November 2019

Snowden, Edward, *Permanent Record* (Macmillan, 2019)

Solon, Olivia, 'Ashamed to Work in Silicon Valley: How Techies Became the New Bankers', *Guardian*, 8 November 2017

Solon, Olivia, '"Data Is a Fingerprint": Why You Aren't as Anonymous as You Think Online', *Guardian*, 13 July 2018

Solon, Olivia, '"Surveillance Society": Has Technology at the US-Mexico Border Gone Too Far?', *Guardian*, 13 June 2018

'Something Doesn't Ad Up About America's Advertising Market', *The Economist*, 18 January 2018

St John, Allen, 'How Facebook Tracks You, Even When You're Not on Facebook', *Consumer Reports*, 11 April 2018

Stanokvic, L., V. Stanokvic, J. Liao and C. Wilson, 'Measuring the Energy Intensity of Domestic Activities From Smart Meter Data', *Applied Energy* 183, 2016

Statt, Nick, 'Facebook CEO Mark Zuckerberg Says the "Future is Private"', *Verge*, 30 April 2019

Statt, Nick, 'How AT&T's Plan to Become the New Facebook Could Be a Privacy Nightmare', *Verge*, 16 July 2018

Statt, Nick, 'Peter Thiel's Controversial Palantir Is Helping Build a Coronavirus Tracking Tool for the Trump Admin', *Verge*, 21 April 2020

Stehr, Nico and Marian T. Adolf, 'Knowledge/Power/Resistance', *Society* 55, 2018

Stokel-Walker, Chris, 'Zoom Security: Take Care With Your Privacy on the Video App', *The Times*, 12 April 2020

Stone, Linda, 'The Connected Life: From Email Apnea to Conscious Computing', *Huffington Post*, 7 May 2012

Strickland, Eliza, 'How IBM Watson Overpromised and Underdelivered on AI Health Care', *IEEE Spectrum*, 2 April 2019

Susskind, Jamie, *Future Politics. Living Together in a World Transformed by Tech* (Oxford University Press, 2018)

Talisse, Robert B., 'Democracy: What's It Good For?', *Philosophers' Magazine* 89, 2020

Tandy-Connor, S., J. Guiltinan, K. Krempely, H. LaDuca, P. Reineke, S. Gutierrez, P. Gray and B. Tippin Davis, 'False-Positive Results Released by Direct-to-Consumer Genetic Tests Highlight the Importance of Clinical Confirmation Testing for Appropriate Patient Care', *Genetics in Medicine* 20, 2018

Tang, Frank, 'China Names 169 People Banned From Taking Flights or Trains Under Social Credit System', *South China Morning Post*, 2 June 2018

Tanner, Adam, *Our Bodies, Our Data. How Companies Make Billions Selling Our Medical Records* (Beacon Press, 2017)

Thompson, Stuart A. and Charlie Warzel, 'Twelve Million Phones, One Dataset, Zero Privacy', *New York Times*, 19 December 2019

Thomson, Amy and Jonathan Browning, 'Peter Thiel's Palantir Is Given Access to U.K. Health Data on Covid-19 Patients', *Bloomberg*, 5 June 2020

Tiku, Nitasha, 'Privacy Groups Claim Online Ads Can Target Abuse Victims', *Wired*, 27 January 2019

Tondo, Lorenzo, 'Scientists Say Mass Tests in Italian Town Have Halted Covid-19 There', *Guardian*, 18 March 2020

Trafton, Anne, 'Artificial Intelligence Yields New Antibiotic', MIT News Office, 20 February 2020

Trump, Kris-Stella, 'Four and a Half Reasons Not to Worry That Cambridge Analytica Skewed the 2016 Election', *Washington Post*, 23 March 2018

Turton, William, 'Why You Should Stop Using Telegram Right Now', *Gizmodo*, 24 June 2016

Tynan, Dan, 'Facebook Says 14m Accounts Had Personal Data Stolen in Recent Breach', *Guardian*, 12 October 2018

'Update Report Into Adtech and Real Time Bidding' (United Kingdom: Information Commissioner's Office, 2019)

Valentino-DeVries, Jennifer, Natasha Singer, Michael H. Keller and Aaron Krolik, 'Your Apps Know Where You Were Last Night, and They're Not Keeping It Secret', *New York Times*, 10 December 2018

Véliz, Carissa, 'Data, Privacy and the Individual' (Madrid: Center for the Governance of Change, IE University, 2020)

Véliz, Carissa, 'Inteligencia artificial: ¿progreso o retroceso?', *El País*, 14 June 2019

Véliz, Carissa, 'Privacy is a Collective Concern', *New Statesman*, 22 October 2019

Véliz, Carissa, 'Why You Might Want to Think Twice About Surrendering Online Privacy for the Sake of Convenience', *The Conversation*, 11 January 2017

Véliz, Carissa and Philipp Grunewald, 'Protecting Data Privacy Is Key to a Smart Energy Future', *Nature Energy* 3, 2018

Victor, Daniel, 'What Are Your Rights if Border Agents Want to Search Your Phone?', *New York Times*, 14 February 2017

Vold, Karina and Jess Whittlestone, 'Privacy, Autonomy, and Personalised Targeting: Rethinking How Personal Data Is Used', in Carissa Véliz (ed.), *Data, Privacy, and the Individual* (Center for the Governance of Change, IE University, 2019)

Waddell, Kaveh, 'A NASA Engineer Was Required To Unlock His Phone At The Border', *Atlantic*, 13 February 2017

Wall, Matthew, '5G: "A Cyber-Attack Could Stop the Country"', BBC News, 25 October 2018

Wallace, Gregory, 'Instead of the Boarding Pass, Bring Your Smile to the Airport', CNN, 18 September 2018

Wang, Echo and Carl O'Donnell, 'Behind Grindr's Doomed Hookup in China, a Data Misstep and Scramble to Make Up', Reuters, 22 May 2019

Wang, L., L. Ding, Z. Liu, L. Sun, L. Chen, R. Jia, X. Dai, J. Cao and J. Ye, 'Automated Identification of Malignancy in Whole-Slide Pathological Images: Identification of Eyelid Malignant Melanoma in Gigapixel Pathological Slides Using Deep Learning', *British Journal of Ophthalmology* 104, 2020

Wang, Orange, 'China's Social Credit System Will Not Lead to Citizens Losing Access to Public Services, Beijing Says', *South China Morning Post*, 19 July 2019

Warzel, Charlie, 'Chinese Hacking Is Alarming. So Are Data Brokers', *New York Times*, 10 February 2020

Warzel, Charlie and Ash Ngu, 'Google's 4,000-Word Privacy Policy Is a Secret History of the Internet', *New York Times*, 10 July 2019

Watson, Gary, 'Moral Agency', *The International Encyclopedia of Ethics* (2013)

Weber, M., *Economy and Society* (Berkeley: University of California Press, 1978)

Weinberg, Gabriel, 'What If We All Just Sold Non-Creepy Advertising?', *New York Times*, 19 June 2019

Weiss, Mark, 'Digiday Research: Most Publishers Don't Benefit From Behavioral Ad Targeting', *Digiday*, 5 June 2019

Whittaker, Zack, 'A Huge Database of Facebook Users' Phone Numbers Found Online', *TechCrunch*, 4 September 2019

Williams, James, *Stand Out of Our Light. Freedom and Resistance in the Attention Economy* (Cambridge: Cambridge University Press, 2018)

Williams, Oscar, 'Palantir's NHS Data Project "may outlive coronavirus crisis"', *New Statesman*, 30 April 2020

Wilson, James H., Paul R. Daugherty and Chase Davenport, 'The Future of AI Will Be About Less Data, Not More', *Harvard Business Review*, 14 January 2019

Wolff, Jonathan, 'The Lure of Fascism', *Aeon*, 14 April 2020

Wolfson, Sam, 'Amazon's Alexa Recorded Private Conversation and Sent it to Random Contact', *Guardian*, 24 May 2018

Wolfson, Sam, 'For My Next Trick: Dynamo's Mission to Bring Back Magic', *Guardian*, 26 April 2020

Wong, Edward, 'How China Uses LinkedIn to Recruit Spies Abroad', *New York Times*, 27 August 2019

Wood, Chris, 'WhatsApp Photo Drug Dealer Caught By "Groundbreaking Work"', *BBC News*, 15 April 2018

Wu, Tim, *The Attention Merchants* (Atlantic Books, 2017)

Wu, Tim, 'Facebook Isn't Just Allowing Lies, It's Prioritizing Them', *New York Times*, 4 November 2019

Wylie, Christopher, *Mindf*ck. Inside Cambridge Analytica's Plot to Break the World* (London: Profile Books, 2019)

Yadron, Danny, 'Silicon Valley Tech Firms Exacerbating Income Inequality, World Bank Warns', *Guardian*, 15 January 2016

Zetter, Kim, 'The NSA Is Targeting Users of Privacy Services, Leaked Code Shows', *Wired*, 3 July 2014

Zittrain, Jonathan, 'Facebook Could Decide an Election Without Anyone Ever Finding Out', *New Statesman*, 3 June 2014

Zittrain, Jonathan, 'How to Exercise the Power You Didn't Ask For', *Harvard Business Review*, 19 September 2018

Zou, James and Londa Schiebinger, 'AI Can Be Sexist and Racist – It's Time to Make It Fair', *Nature* 559, 2018

Zuboff, Shoshana, *The Age of Surveillance Capitalism* (London: Profile Books, 2019)

INDEX

CARISSA VÉLIZ is an Associate Professor at the Faculty of Philosophy and the Institute for Ethics in AI, and a Tutorial Fellow at Hertford College, University of Oxford. She works on privacy, technology, moral and political philosophy, and public policy.

Véliz has published articles in the *Guardian*, the *New York Times*, *New Statesman*, the *Independent*, *Slate*, and *El País*. Her academic work has been published in the *Harvard Business Review*, *Nature Electronics*, *Nature Energy* and *Bioethics*, among other journals. She is the editor of the *Oxford Handbook of Digital Ethics*.

@carissaveliz